AP* SUCCESS
Calculus AB/BC

4th edition

Joan Van Galbek
Lawrence Trivieri
Lalit A. Ahuja
Jessica Polito
Joan Marie Rosebush

THOMSON
PETERSON'S

Australia • Canada • Mexico • Singapore • Spain • United Kingdom • United States

About The Thomson Corporation and Peterson's

With revenues of US$7.2 billion, The Thomson Corporation (www.thomson.com) is a leading global provider of integrated information solutions for business, education, and professional customers. Its Learning businesses and brands (www.thomsonlearning.com) serve the needs of individuals, learning institutions, and corporations with products and services for both traditional and distributed learning.

Peterson's, part of The Thomson Corporation, is one of the nation's most respected providers of lifelong learning online resources, software, reference guides, and books. The Education Supersite℠ at www.petersons.com—the Internet's most heavily traveled education resource—has searchable databases and interactive tools for contacting U.S.-accredited institutions and programs. In addition, Peterson's serves more than 105 million education consumers annually.

Editorial Development: American BookWorks Corporation
Contributing Editors: Jesse W. Byrne, John W. Watson, David A. Miller
Art and Design: James Snyder

For more information, contact Peterson's, 2000 Lenox Drive, Lawrenceville, NJ 08648; 800-338-3282; or find us on the World Wide Web at www.petersons.com/about.

COPYRIGHT © 2003 Peterson's, a division of Thomson Learning, Inc.

Previous editions © 2000, 2001

Thomson Learning™ is a trademark used herein under license.

ALL RIGHTS RESERVED. No part of this work covered by the copyright herein may be reproduced or used in any form or by any means—graphic, electronic, or mechanical, including photocopying, recording, taping, Web distribution, or information storage and retrieval systems—without the prior written permission of the publisher.

For permission to use material from this text or product, contact us by
Phone: 800-730-2214
Fax: 800-730-2215
Web: www.thomsonrights.com

ISBN 0-7689-0980-5

Printed in the United States of America

10 9 8 7 6 5 4 3 2 1   05 04 03

CONTENTS

Introduction 1

RED ALERT

About the Tests 3
AP Calculus Study Plan 6
    The 9-Week Plan 6
    The 18-Week Plan 9
    Using the Calculator 10
Diagnostic Test, AB 13
Diagnostic Test, BC 23
Answers and Explanations 33

CALCULUS REVIEW

Part I: Functions, Graphs, and Limits 45
Part II: Derivatives 83
Part III: Integrals 131
Part IV: Polynomial Approximations and Series 183
AP Calculus AB Practice Test 1 217
AP Calculus AB Practice Test 2 229
AP Calculus BC Practice Test 1 243
AP Calculus BC Practice Test 2 263
Answers and Explanations 285

INTRODUCTION

ABOUT THIS BOOK

The AP Calculus Exam is the culmination of all of your hard work in your AP Calculus class. This book will help you prepare for one of the two AP exams in calculus offered by the College Board. It covers all of the topics required for both Calculus AB and Calculus BC. As you probably know, the Calculus BC exam covers all of the material included in the AB exam, as well as additional topics. In order to take the BC examination, you will need a thorough understanding of the material covered in Calculus AB.

This book is divided into three major sections. The first part includes two diagnostic exams—one for Calculus AB and one for Calculus BC. They are not as extensive as the actual, full-length exams, but will test your overall knowledge of the material and will give you a quick estimate of what subjects still require your attention. For the best results, take the appropriate tests under simulated test-taking conditions—find a quiet place to work, continue through the entire exam without taking a break, and time yourself. By evaluating the results of the exam in terms of the number of questions you answered correctly as well as the time it took you to complete the test, you should be able to determine how prepared you are for the final exam. We highly recommend that you take *both* the AB and BC diagnostic tests, even if your intention is to take only one exam, to get a stronger sense of your preparedness.

The second part of the book contains review material. Please keep in mind that this is not a textbook but rather an approach to brushing up your skills prior to taking the exams. Throughout the book we have used the "solved problem" approach as a study aid. Each topic covered in the exam (and in your calculus classroom work) is presented here in a brief manner; usually, you are given a few sentences to refresh your memory about the topic. Then, the topic is followed by several problems along with their solutions. With this method, you are shown exactly *how* to solve the material, step by step. It has an even greater impact if you cover up the solutions as you work and try to answer the questions on your own. After you're satisfied with your answers, check the solutions to see how well you did.

Finally, the third part of this book presents four full-length simulated calculus exams—two for the AB exam and two for the BC exam. These tests, like the diagnostic tests, should be taken under simulated test-taking conditions. Not only is accuracy important, but so is timing. You should set aside slightly more than three hours for each test, since this is the time allotted on the actual examination. Work in a quiet place, work quickly, and work accurately. We again recommend that you take all four tests, even if you are only taking the Calculus AB exam.

INTRODUCTION

Set aside separate days for each test, and when you have finished taking a test, take a break. Then go back and check your answers and review the solutions. Find out which questions gave you the most difficulty so you can concentrate on them before the actual exam. Try to figure out why you answered the question incorrectly. Were your calculations incorrect, or did you just not know the material? When in doubt, you can always ask your teacher for additional help.

RED ALERT

ABOUT THE TESTS

Both AP Calculus exams consist of two sections. Section I has two parts consisting of 45 multiple-choice questions and lasts 105 minutes. Part A contains 28 questions, and you may not use a calculator. Part B contains 17 questions, and for some of these questions a graphing calculator will be necessary or advantageous. You are given 55 minutes to complete Part A, and 50 minutes to complete Part B. In terms of scoring, one fourth of the number of questions answered incorrectly will be subtracted from the number you answered correctly. Therefore, we suggest that if you can't figure out the answer quickly and easily, rather than just answering a question with the hope that you'll get it right, try to make only educated guesses. If you just don't know the answer, go on to the next question. However, if you can narrow down the choices to only two out of five, then it's worth taking a chance.

Section II consists of two parts, Part A and Part B. Each has three free-response questions for which a graphing calculator is required. This section will take 45 minutes. Part B also has three free-response questions, but you will not be allowed to use a calculator. This section will also take 45 minutes. During Part B, you will still be allowed to work on Part A questions, but you will no longer be permitted to use the calculator.

TEST GUIDELINES

There are a few guidelines that will be helpful to you:

1. Write clearly. If those scoring the test cannot read or interpret your answer, you'll lose credit. It's smarter to cross out your work, which is faster and clearer, than erasing it.

2. You should show all of your work and calculations because you will be graded on the completeness of your methods, as well as the accuracy of your answer. Answers must show supporting work.

3. You should clearly identify functions, graphs, tables, or any other objects you use to justify your answers.

4. Your work should be expressed in standard mathematical notation, rather than calculator syntax. This (fnInt $x2, x, 1, 4$) is not a substitute for $\int_1^4 x^2 dx$.

RED 3 **ALERT**

5. Numeric and algebraic answers do not need to be simplified. Decimal approximations should be correct to three places.

6. The domain of a function f is assumed to be the set of all real numbers x for which $f(x)$ is a real number.

A complete understanding of the guidelines as well as the instructions within each test will be very important as you take the test. It will save you a lot of time if you're familiar with these, since you won't have to waste time reading and rereading directions.

Using Calculators

Students taking the Calculus AB and Calculus BC examinations are required to bring a graphing calculator to the test. Calculators should possess the following capabilities:

1. Plot the graph of a function within an arbitrary viewing window

2. Find the zeros of functions (solve equations numerically)

3. Numerically calculate the derivative of a function

4. Numerically calculate the value of a definite integral

Your teacher will tell you which calculators are approved. If you wish to use a calculator that is not on the approved list, your AP teacher must contact ETS prior to April 1 of the testing year to receive written permission to bring it to the exam.

Topical Outlines

The topical outline for Calculus BC includes all Calculus AB topics. Additional BC topics are marked with an asterisk (*).

I. Functions, Graphs, and Limits
   Analysis of graphs
   Limits of functions (including one-sided limits)
   Asymptotic and unbounded behavior
   Continuity as a property of functions
   * Parametric, polar, and vector functions

II. Derivatives
   Concept of the derivative
   Derivative at a point
   Derivative as a function
   Second derivatives
   Applications of derivatives
   Computation of derivatives

III. Integrals
   Riemann sums
   Interpretations and properties of definite integrals
   Applications of integrals
   Fundamental Theorem of Calculus
   Techniques of antidifferentiation
   Applications of antidifferentiation

*IV. Polynomial Approximations and Series
   * Concept of series
   * Series of constants
   * Taylor series

Scoring

The AP examinations are scored on a five-point scale, as follows:

5 Extremely well-qualified
4 Well qualified
3 Qualified
2 Possibly qualified
1 No recommendation

Most colleges and universities will accept a score of 3 or better. Some, however, require higher scores, so it's important to know the policies of the schools to which you are applying or have been accepted. The first part of the test is machine scored, and as we stated earlier in this section, one fourth of the number of incorrect answers is subtracted from the number of correct answers. The purpose of this is to try to eliminate wild guesses. The second section of the test is graded by college and high school teachers. These "readers" have sample solutions to each problem, and they are given scoring scales and point distributions so that they can allow partial credit for your answers, if necessary. The scores are combined and then converted into one of the five grades above.

In the last few years, the scoring of the BC Calculus test has changed and now provides a subscore grade for the AB Calculus portion, which makes up about 60 percent of the BC exam, in addition to a Calculus BC grade.

Now that you have a general idea about the test, it's time to start working on the diagnostics. Work quickly and accurately, and try to remember the directions so that you won't have to take time reading them again on the actual exam. The key to success on any test, including this one, is practice, practice, practice.

AP CALCULUS STUDY PLAN

So you're getting ready to take the AP Calculus Test. How much time do you have? We offer you these different study plans to help maximize your time and studying. The first is the 9-Week Plan, which involves concentrated studying and a focus on the sample test results. The second is the 18-Week Plan, or Semester Plan, favored by schools; and finally, the Panic Plan, for those of you who have only a few weeks to prepare. Obviously, the more time you have to prepare, the easier it will be to review all of the material—and you will find yourself somewhat more relaxed when taking the actual exam.

These plans are not set in stone. You can feel free to modify them to suit your needs and your own study habits. But start immediately. The more you study and review the questions, the better your results will be.

THE 9-WEEK PLAN—2 LESSONS PER WEEK

Week 1

Lesson 1   Diagnostic Test 1

Take the entire AB diagnostic test in one sitting. There are two major parts of the exam—multiple choice and free response. Normally, the actual complete test will take a total of 180 minutes (each part is 90 minutes), but the diagnostic tests are half-length tests. We suggest that even if you're taking the BC exam, you should spend time on this test to help you review material that most likely will appear on the BC exam.

Diagnostic Answers

Once you have completed the test, it is necessary to check all of your answers and read through the explanations. This may take quite a bit of time, as will all of the tests, but it will enable you to select those subject areas that you should focus on and spend the most amount of time studying.

Lesson 2   Diagnostic Test 2

Take the entire BC diagnostic test in one sitting—even if you are planning to take only the AB exam. It will provide you with an excellent way to test how much you actually know, since a good percentage of the questions on the BC test will be at the AB level.

Diagnostic Answers

Once you have completed the second test, check all of your answers and read through the explanations. Now you should have a greater understanding of what areas will need the most work.

Week 2

Lesson 1   Chapter 1, Functions, Graphs, and Limits

Read through about half of the chapter and make notes throughout. The chapter is filled with solved problems and you should focus on the suggested approaches to solving these problems.

Lesson 2 **Chapter 1, Completion**
Complete your reading of this chapter. At this point you should have a clear understanding of the material that will be covered on the test in this area. The book covers both AB and BC material, so you will have had an in-depth review of this material.

WEEK 3

Lesson 1 **Chapter 2, Derivatives**
Study the first half of this chapter and work through the solved problems. We suggest that you work on the problems *before* checking the answers to make sure you really understand the material.

Lesson 2 **Chapter 2, Completion**
Complete your reading of this chapter. At this point you should have a clear understanding of the material that will be covered on the test in this area.

WEEK 4

Lesson 1 **Chapter 3, Integrals**
Continue with your AP Calculus preparation by studying the first half of this chapter and working through the solved problems. Whenever you run into trouble, consult your textbook or your teacher for clarification.

Lesson 2 **Chapter 3, Completion**
Complete your reading of this chapter. At this point you should have a clear understanding of the topic of integrals and the material that will be covered on the test in this area.

WEEK 5

Lesson 1 **Chapter 4, Polynomial Approximations and Series**
This is the final chapter of the review material. Read and study the first half of this chapter and work through the solved problems. Whenever you run into trouble, consult your textbook or your teacher for clarification.

Lesson 2 **Chapter 3, Completion**
Complete your reading of this chapter. At this point you should have a clear understanding of the topic of polynomial approximations and series and the material that will be covered on the test in this area.

WEEK 6

Lesson 1 **AP Calculus AB Practice Test #1**
Take this test and answer all of the questions you can and then guess at those you don't know. Circle those questions that you guessed at so you can zero in on those specific answers—and don't delude yourself into thinking that you really knew those answers in the first place.

Lesson 2 AP Calculus AB Practice Test #1, Answers
Check all of your answers to both parts of the test. Keep track of the specific questions that are still giving you problems. It is understandable to make mistakes or just not know an answer. However, what is more important is to understand the type of question and to make sure that you can review the topics that you don't understand.

Week 7

Lesson 1 AP Calculus AB Practice Test #2
Take this test and answer all of the questions you can and then guess at those you don't know. Circle those questions that you guessed at so that you can zero in on those specific answers—and don't delude yourself into thinking that you really knew those answers in the first place.

Lesson 2 AP Calculus AB Practice Test #2, Answers
Check all of your answers to both parts of the test.

Week 8

Lesson 1 AP Calculus BC Practice Test #1
Take this test and answer all of the questions you can, and then guess at those you don't know. Circle those questions that you guessed at, although by now you should have a better understanding of material.

Lesson 2 AP Calculus BC Practice Test #1, Answers
Check all of your answers to both parts of the test.

Week 9

Lesson 1 AP Calculus BC Practice Test #2
Take this test even if you're only planning to take the AB exam, and answer all of the questions you can, and then guess at those you don't know. Circle those questions that you guessed at. If you're only taking the AB exam, you may have more difficulty with this test.

Lesson 2 AP Calculus BC Practice Test #2, Answers
Check all of your answers to both parts of the test.

THE 18-WEEK PLAN—1 LESSON PER WEEK

For those of you who have the extra time, the 18-Week Plan will enable you to better utilize your study time. You will be able to spread out your plan into one lesson a week. This plan is ideal because you are not under any pressure and you can take more time to review the material in each of the chapters. You will also have enough time to double-check the answers to questions that might have given you problems. The basis for all test success is practice, practice, practice.

THE PANIC PLAN

While we hope you don't fall into this category, not everyone has the luxury of extra time to prepare for the AP Calculus Exam. However, perhaps we can offer you a few helpful hints to get your though this period.

1. Read through the official AP Calculus bulletin and this book and memorize the directions. One way of saving time on this, or any, test, is to be familiar with the directions in order to maximize the time you have to work on the questions.

2. Read the introduction to this book. It will be helpful in preparing for the test and will give you an understanding of what you can expect on the exam and how much time you will have to complete both sections of the test.

3. Take the diagnostic tests as well as the practice tests. Depending upon how much time you have, focus only on those tests that are applicable to the version of the test you are planning to take. If, for example, you are taking the BC Calculus test, take only those practice tests. If you have time afterward, then you can take the AB version.

4. Focus whatever time you have left on those specific areas of the test that gave you the most difficulty when you took the practice tests.

Whatever time you have before the exam, keep in mind that the more you practice, the better you will do on the final exam.

USING THE CALCULATOR

The AP Calculus Development Committee adopted the use of calculators based on the strong recommendations of a variety of professional mathematics organizations, such as the National Council of Teachers of Mathematics, the Mathematical Association of America, and the Mathematical Sciences Education Board of the National Academy of Sciences.

As such, part of the AP Calculus Examination will require the use of a graphing calculator. It is also likely that you will spend some time in the classroom working on the use of these calculators.

CALCULATOR REQUIREMENTS

The only requirements regarding a calculator are that it has the ability to:

1. produce the graph of a function within an arbitrary viewing window,

2. find the zeros of a function,

3. compute the derivative of a function numerically, and

4. compute definite integrals numerically.

These abilities can either be built into the calculator or can be programmed into the calculator before the examination. Computers, non-graphing scientific calculators, devices with a typewriter-type keyboard, and electronic writing pads are not allowed.

Make sure that you use an approved calculator, since the test administrator will be checking all calculators. It *is* permissible to bring up to two graphing calculators with you to the test, but to be safe, make sure you bring extra batteries. You may not share your calculator with another student and you cannot use your calculator to take any information out of the testing room. To do so would invalidate your grade.

THE USE OF THE GRAPHING CALCULATOR ON THE FREE-RESPONSE SECTION

The use of graphing calculators on the free-response section of the exam will affect both the way some questions may be asked as well as how you answer them. You will always be expected to show your work, not just the final answer. However, for any solutions you obtain using one of the four required capabilities listed above, you will be required to include your setup (a definite integral, an equation, or a derivative) that leads to the solution, along with the results you found using your calculator. If, however, you obtain a result with any capability other than one of the required four functions, you must show the mathematical steps leading to your answer. Without these explanations, you will not receive full credit.

A decimal answer must be correct to three decimal places unless otherwise indicated. Be careful about rounding values before calculations are concluded.

Approved Calculators

There is a long list of approved calculator models for the AP Calculus exam, but since the list is always growing, we will not include the model numbers here. Among the brands that are acceptable are the following:

Casio
Hewlett-Packard
Radio Shack
Sharp
Texas Instruments

There are dozens of acceptable models within these brands, so make sure you have an approved list. To obtain the latest information about which specific models are approved for the examination, you can ask your teacher, contact the College Board, or visit their Web site at www.collegeboard.com, and follow the links to the AP Calculus Exam.

If you plan to use a machine that is not on the approved list, you must have your teacher contact ETS prior to April 1 of the year in which you plan to take the test in order to receive *written* permission for your calculator.

ADVANCED PLACEMENT CALCULUS AB

DIAGNOSTIC TEST

Time	Number of Questions
27 Minutes	14

A CALCULATOR MAY NOT BE USED ON THIS PART OF THE EXAMINATION

Section I, Part A

Directions: Solve each of the following problems using the available space for scratchwork. After examining the form of the choices, decide which is the best of the choices given and fill in the corresponding oval on the answer sheet. No credit will be given for anything written in the test book. Do not spend too much time on any one problem.

<u>In this test:</u> Unless otherwise specified, the domain of a function f is assumed to be the set of all real numbers x for which $f(x)$ is a real number.

1. $\int x^3 - 3x^2 + 4x - 1 \, dx$ is equal to

    (A) $\dfrac{x^4}{4} - x^3 + 2x^2 - x + C$

    (B) $x^4 - 3x^3 + 4x^2 - x + C$

    (C) $\dfrac{x^4}{4} - x^3 + 4x^2 - x + C$

    (D) $3x^2 - 6x + 4 + C$

    (E) $\dfrac{x^3}{3} - x^2 + 2x - 1 + C$

2. On what interval is the function $f(x) = x^3 + 3x^2 - 24x + 1$ decreasing?

    (A) $(-\infty, -4)$
    (B) $(-4, 2)$
    (C) $(2, \infty)$
    (D) $(0, 4)$
    (E) $(-4, -2)$

3. $\int_0^{\pi/6} \sin 3x \, dx$ is equal to

(A) $\frac{1}{3}$
(B) -3
(C) 3
(D) $-\frac{1}{3}$
(E) 0

4. If $u = \sin 5x$ and $y = u^2$, what is dy/dx?

(A) $2 \sin 5x$
(B) $10 \cos 5x$
(C) $5 \cos^2 5x$
(D) $5 \sin^2 5x$
(E) $10(\sin 5x)(\cos 5x)$

5. What is the absolute minimum value of the function $f(x) = x - 2 \ln x$?

(A) $2 + 2 \ln 2$
(B) $\frac{1}{2}$
(C) $2 - 2 \ln 2$
(D) 2
(E) $\frac{1}{2 \ln 2}$

6. What is $\lim_{h \to 0} \frac{\ln(1+h)}{h}$?

(A) 0
(B) 1
(C) -1
(D) nonexistent
(E) none of the above

7. Which of the following limits is undefined?

   I. $\lim_{x \to 0} \dfrac{\sin 2x}{x}$

   II. $\lim_{x \to 2} \dfrac{x^2 + x - 6}{x - 2}$

   III. $\lim_{x \to 1} \dfrac{x^2 + 4x + 1}{x - 1}$

   (A) I only
   (B) III only
   (C) I and II only
   (D) I and III only
   (E) II and III only

8. If $f(x) = \dfrac{e^{3x}}{\sin x}$ then $f'(x) =$

   (A) 0

   (B) $\dfrac{e^{3x}(\cos x - 3\sin x)}{\sin^2 x}$

   (C) $\dfrac{e^{3x}(3\cos x - \sin x)}{\sin^2 x}$

   (D) $\dfrac{e^{3x}(\sin x - 3\cos x)}{\sin^2 x}$

   (E) $\dfrac{e^{3x}(3\sin x - \cos x)}{\sin^2 x}$

9. The solution to the differential equation $\dfrac{dy}{dx} = \dfrac{y^2}{x}$ which passes through the point $(1, -1)$ is

   (A) $y = -\dfrac{1}{1 + \ln|x|}$

   (B) $y = -x^2$

   (C) $y = -\dfrac{\ln|x| + 1}{x}$

   (D) $y = 3x^2 - 4$

   (E) none of the above

10. Let $f(x)$ be a function with a continuous derivative on the interval (1, 3) such that $f(1) = 2$ and $f(3) = -4$. Which of the following must be true for some a in (1, 3)?

   (A) $f'(a) = 6$
   (B) $f'(a) = 3$
   (C) $f'(a) = 0$
   (D) $f'(a) = -3$
   (E) $f'(a) = -6$

11. Let $f(x)$ be a function with $\int_1^3 f(x)dx = 2$ and $\int_3^7 f(x)dx = 1$. Let $g(x)$ be a function such that $\int_1^7 g(x)dx = -3$.

   What is $\int_1^7 2f(x) - g(x)dx$?

   (A) 0
   (B) 1
   (C) 3
   (D) 6
   (E) 9

The graph of $f'(x)$ shown is for problems 12 and 13

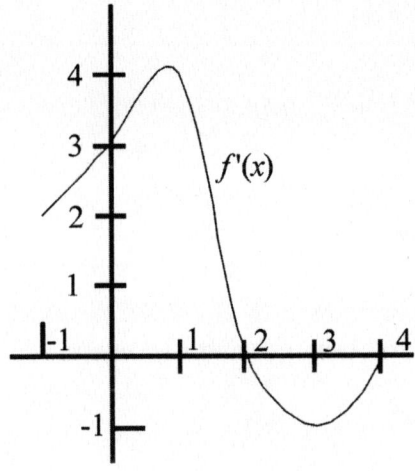

12. $f(x)$ has a local minimum at

   (A) $x = 1$ only
   (B) $x = 2$ only
   (C) $x = 4$ only
   (D) $x = 2$ and $x = 4$ only
   (E) $x = 1$ and $x = 4$

13. On which of the following intervals is $f(x)$ concave up?

   (A) $(-1, 1)$ and $(1, 2)$
   (B) $(-1, 1)$ and $(3, 4)$
   (C) $(1, 2)$ and $(3, 4)$
   (D) $(-1, 0)$ and $(1, 2)$
   (E) $(-1, 2)$ and $(3, 4)$

14. A particle is moving along the x-axis. Its position at time $t > 0$ is $\dfrac{2t}{1+t^2}$. When is it stationary?

   (A) $t = 0$
   (B) $t = 1$
   (C) $t = 2$
   (D) $t = 3$
   (E) $t = 4$

ADVANCED PLACEMENT CALCULUS AB

Time	Number of Questions
25 Minutes	9

A GRAPHING CALCULATOR IS REQUIRED FOR SOME QUESTIONS ON THIS PART OF THE EXAMINATION

Section I, Part B

Directions: Solve each of the following problems using the available space for scratchwork. After examining the form of the choices, decide which is the best of the choices given and fill in the corresponding oval on the answer sheet. No credit will be given for anything written in the test book. Do not spend too much time on any one problem.

<u>In this test:</u> The exact numerical value of the correct answer does not always appear among the choices given. When this happens, select from among the choices the number that best approximates the exact numerical value.

<u>Note:</u> Unless otherwise specified, the domain of a function f is assumed to be the set of all real numbers x for which $f(x)$ is a real number.

---

15. If $f(1) = 5.4$ and $f'(1) = 1.5$, which of the following is the best approximation for $f(1.2)$?

    (A) 0.3
    (B) 1.5
    (C) 5.6
    (D) 5.7
    (E) 6.9

---

16. What is the average value of the function $f(x) = \dfrac{x}{x^2 + 1}$ on the interval $(0, 2)$?

    (A) 0.38
    (B) 0.40
    (C) 0.42
    (D) 0.44
    (E) 0.50

17. What is the slope of the curve $f(x) = 3x\sqrt{2x^2 + 1}$ at the point $x = 2$?

    (A) 1
    (B) 6
    (C) 17
    (D) 18
    (E) 27

18. Two sailboats set out from the same dock at noon. One is sailing north at a speed of 7 mph, and the other is sailing west at a speed of 3 mph. How many mph is the distance between them changing at 2:00?

    (A) 7.49
    (B) 7.54
    (C) 7.62
    (D) 7.78
    (E) 7.83

19. What is the volume of the solid obtained by rotating the region under the graph $y = \sqrt{x^3 + 1}$ between $x = 1$ and $x = 2$, around the x-axis?

    (A) $\pi \int_{1}^{2} x^3 + 1 \, dx$

    (B) $\pi \int_{1}^{2} 2\sqrt{x^3 + 1} \, dx$

    (C) $\pi \int_{0}^{\ln(2)} x^3 + 1 \, dx$

    (D) $\pi \int_{0}^{\ln(2)} 2\sqrt{x^3 + 1} \, dx$

    (E) $\pi \int_{1}^{\ln(2)} \sqrt{x^3 + 1} \, dx$

20. A bacteria colony grows at a rate proportional to its size. The initial population was 100; in 2 hours, the population was 200. What was the population after 3 hours?

    (A) 275
    (B) 283
    (C) 294
    (D) 300
    (E) 332

21. The graph of function $y = x^3 + 4x^2 - 8x + 3\cos(x)$ has a point of inflection at

    (A) −2.31
    (B) −1.97
    (C) −1.11
    (D) −1.04
    (E) −0.87

22. What is $\lim_{x \to \infty} \dfrac{3x^2 + 2x + 2}{4x^2 + x + 5}$?

    (A) 0

    (B) $-\dfrac{3}{4}$

    (C) $\dfrac{3}{4}$

    (D) $\dfrac{2}{5}$

    (E) ∞

23. Let $y = 2x^2 - 1$. What is the minimum value of $x^2 y$?

    (A) 0

    (B) $-\dfrac{1}{8}$

    (C) $-\dfrac{1}{4}$

    (D) $-\dfrac{1}{2}$

    (E) −1

DIAGNOSTIC TEST, SECTION II

Time	Number of Questions
45 Minutes	3

A GRAPHING CALCULATOR IS REQUIRED FOR SOME QUESTIONS ON THIS PART OF THE EXAMINATION

SECTION II

Directions: Solve each of the following problems using the available space for scratchwork. After examining the form of the choices, decide which is the best of the choices given and fill in the corresponding oval on the answer sheet. No credit will be given for anything written in the test book. Do not spend too much time on any one problem.

In this test: Show all your work. You will be graded on the correctness and completeness of your methods as well as the accuracy of your final answers. Correct answers without supporting work may not receive credit.

Justification requires that you give mathematical (noncalculator) reasons.

You are permitted to use your calculator to solve an equation, find the derivative of a function at a point, or calculate the value of a definite integral. However, you must clearly indicate the setup of your program, namely the equation, function, or integral you are using. If you use other built-in features or programs, you must show the mathematical steps necessary to produce your results.

Note: Unless otherwise specified, answers (numeric or algebraic) need not be simplified. If your answer is given as a decimal approximation, it should be correct to three places after the decimal point.

---

1. Let f be the function given by $f(x) = (x-1)/e^{x^2}$

    (A) Find $\lim_{x \to \infty} f(x)$.

    (B) When does $f(x)$ have an absolute maximum? Justify that your answer is an absolute maximum.

    (C) When does the function $g(x) = \int_0^x f(t)dt$ have an absolute minimum? Justify that your answer is an absolute minimum.

2. The velocity of a particle moving along the x-axis is given in the following table and graph:

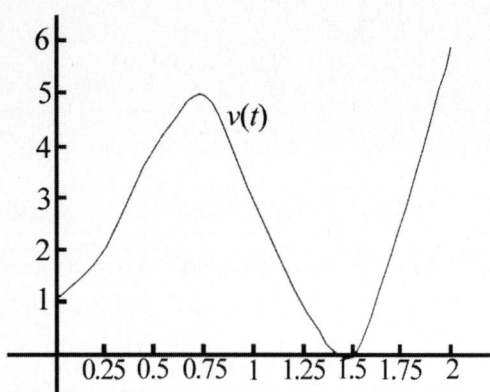

t	v(t)
0	1
0.25	2
0.5	4
0.75	5
1	3
1.25	1
1.5	0
1.75	2
2	5

(A) During which intervals is the acceleration of the particle positive? Justify your answer.
(B) Approximate the acceleration of the particle when $t = 1$.
(C) Determine precisely, if possible, and if not, approximate the average velocity of the particle, and the average acceleration, as t goes from 0 to 2.

3. Let R be the region in the plane bounded by the curve $y = -x^2 + 4x - 3$ and the line $y = x - 1$.

(A) For what x value is the distance from the top to the bottom of the region R maximal?
(B) Find the area of the region R.
(C) Find the volume of the solid generated when R is revolved around the y-axis.

ADVANCED PLACEMENT CALCULUS BC

DIAGNOSTIC TEST

Time	Number of Questions
26 Minutes	14

A CALCULATOR MAY NOT BE USED ON THIS PART OF THE EXAMINATION

Section I, Part A

Directions: Solve each of the following problems using the available space for scratchwork. After examining the form of the choices, decide which is the best of the choices given and fill in the corresponding oval on the answer sheet. No credit will be given for anything written in the test book. Do not spend too much time on any one problem.

In this test: Unless otherwise specified, the domain of a function f is assumed to be the set of all real numbers x for which $f(x)$ is a real number.

---

1. If $x = t^3$ and $y = \ln(1 - t)$, what is dy/dx?

   (A) $-\dfrac{1}{3t^2(1-t)}$

   (B) $\dfrac{1}{t^2(1-t)}$

   (C) $t^2(1-t)$

   (D) $\dfrac{t^2}{(1-t)}$

   (E) $-t^2(1-t)$

---

2. If $f(x) = (x+1)^{3/2} - e^{x^2-9}$, what is $f'(3)$?

   (A) -5
   (B) -3
   (C) 0
   (D) 1
   (E) 3

3. What is the slope of the tangent line to the curve $y = \ln(x^2 + 1)$ when $x = 3$?

(A) $\dfrac{1}{10}$

(B) $\dfrac{3}{10}$

(C) $\dfrac{1}{5}$

(D) $\dfrac{1}{15}$

(E) $\dfrac{3}{5}$

4. What is $\int \dfrac{\ln x}{\sqrt{x}} dx$

(A) $2\sqrt{x} \ln x + C$

(B) $\sqrt{x}(2\ln x - 2) + C$

(C) $2(\ln x - \sqrt{x}) + c$

(D) $2\sqrt{x}(\ln x - 2) + C$

(E) $2\sqrt{x}(\ln + 2) + C$

5. What is $\lim\limits_{x \to \infty} \dfrac{3\sqrt{x^4 - 3x}}{2x^2 + \cos x}$

(A) ∞

(B) undefined

(C) $\dfrac{3}{2}$

(D) 0

(E) 6

6. The first 4 terms of the Taylor series around 0 of $\ln(1 + x^2)$ are

   (A) $x^2 - \dfrac{x^4}{2} + \dfrac{x^6}{3} - \dfrac{x^8}{4}$

   (B) $x^2 + \dfrac{x^4}{2} + \dfrac{x^6}{3} + \dfrac{x^8}{4}$

   (C) $x - \dfrac{x^2}{2} + \dfrac{x^3}{3} - \dfrac{x^4}{4}$

   (D) $x^2 - \dfrac{x^4}{2!} + \dfrac{x^6}{3!} - \dfrac{x^8}{4!}$

   (E) $-x^2 - \dfrac{x^4}{2!} - \dfrac{x^6}{3!} - \dfrac{x^8}{4!}$

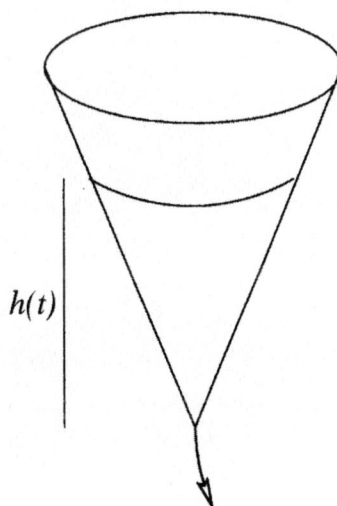

7. Water is pouring out of the base of a conical funnel at a constant rate; refer to the above diagram. Let $h(t)$ be the height of the water in the funnel. Which of the following is negative?

   (I) $h(t)$
   (II) $h'(t)$
   (III) $h''(t)$

   (A) (I) only
   (B) (II) only
   (C) (III) only
   (D) (II) and (III) only
   (E) (I), (II), and (III)

8. What is $\int_1^3 x\sqrt{x^2-1}\,dx$?

   (A) $\dfrac{8^{1/2}}{3}$

   (B) $3^{1/2}$

   (C) $3^{3/2}$

   (D) $8^{3/2}$

   (E) $\dfrac{8^{3/2}}{3}$

9. A function $f(x)$ is equal to $\dfrac{x^2-6x+9}{x-3}$ for all $x > 0$ except $x = 3$. In order for the function to be continuous at $x = 3$, what must the value of $f(3)$ be?

   (A) 5
   (B) 4
   (C) 2
   (D) 1
   (E) 0

10. Which of the following guarantee that $\sum_{n=0}^{\infty} f(n)$ diverges?

    (I) $\lim\limits_{n \to \infty} \dfrac{f(n+1)}{f(n)} = 3$

    (II) $\int_2^{\infty} f(x)\,dx = 5$

    (III) $f(x) > \dfrac{1}{x}$

    (A) (I) only
    (B) (III) only
    (C) (I) and (III) only
    (D) (I) and (II) only
    (E) (I), (II), and (III)

11. For what values of x does the function $5x^2 + 2xy + y^2 = 16$ have a vertical tangent line?

    (A) $x = 2$, only
    (B) $x = 2$ and $x = 0$, only
    (C) $x = 0$, $x = -1$, and $x = -2$, only
    (D) $x = 2$ and $x = -2$, only
    (E) Never

12. If $f(1) = 3$ and $f'(1) = -1$, an estimate of $f(0.95)$ is

    (A) 3.95
    (B) 3.05
    (C) 2.95
    (D) 2.5
    (E) 2.05

13. Let $f(x)$ be a function with a continuous first and second derivative on the interval $(0, 5)$ such that $f'(0) = 4$, $f'(1) = 2$, $f'(2) = -2$, $f'(3) = -1$, $f'(4) = 5$. Which of the following must be true about $f(x)$?

    (I) $f(x)$ has a point of inflection between $x = 1$ and $x = 3$
    (II) $f(x)$ has a point of inflection between $x = 2$ and $x = 4$
    (III) $f(x)$ has a point of inflection between $x = 0$ and $x = 2$

    (A) I only
    (B) II only
    (C) I and II only
    (D) I and III only
    (E) I, II, and III

14. Which of the following series diverge?

    (A) $\sum_{n=1}^{\infty} \dfrac{n}{3n^4 - 1}$

    (B) $\sum_{n=0}^{\infty} \dfrac{n^5 - 1}{e^{2n}}$

    (C) $\sum_{n=0}^{\infty} \dfrac{(-1)^n \cos n}{n^3 + 1}$

    (D) All of the above
    (E) None of the above

ADVANCED PLACEMENT CALCULUS BC

Time	Number of Questions
25 Minutes	9

A GRAPHING CALCULATOR IS REQUIRED FOR SOME QUESTIONS ON THIS PART OF THE EXAMINATION

SECTION I, PART B

Directions: Solve each of the following problems using the available space for scratchwork. After examining the form of the choices, decide which is the best of the choices given and fill in the corresponding oval on the answer sheet. No credit will be given for anything written in the test book. Do not spend too much time on any one problem.

In this test: The exact numerical value of the correct answer does not always appear among the choices given. When this happens, select from among the choices the number that best approximates the exact numerical value.

15. What is the arc length of the curve given by $x = e^t$, $y = t$ as t goes from 0 to 3?

   (A) 19.52
   (B) 20.13
   (C) 21.43
   (D) 23.84
   (E) 25.62

16. What is the radius of convergence of the series $\sum_{n=0}^{\infty} \frac{(x-5)^n}{3^n}$?

   (A) 0
   (B) 1
   (C) 3
   (D) 5
   (E) ∞

17. The position of a partical on the x-axis is given by cos(3t) − ln(4t), where t is the time in seconds. How many times is the acceleration of the partical equal to 0 in the first three seconds?

    (A) 0
    (B) 1
    (C) 2
    (D) 3
    (E) 4

18. You have a sample of five grams of a radioactive substance. It has a half-life of 4 days. How long will it be before one gram remains of the substance?

    (A) 9 days
    (B) 9.288 days
    (C) 9.785 days
    (D) 10 days
    (E) 10.345 days

19. What is the area inside the curve $r = 3 \sin \theta$?

    (A) 3.15
    (B) 5.56
    (C) 6.28
    (D) 7.07
    (E) 14.14

20. Let $f(x) = e^{1-x^2}$. Which of the following are true?

    (I) $\int_1^x f(t)dt$ is always increasing
    (II) $f(x)$ is always increasing
    (III) $f'(x)$ is always increasing

    (A) (I) only
    (B) (III) only
    (C) (I) and (III) only
    (D) (I) and (II) only
    (E) (I), (II), and (III)

21. Which of the following improper integrals diverge?

(A) $\int_1^3 (2x-2)^{-1/3} dx$

(B) $\int_{-\infty}^3 x^2 e^{x^3} dx$

(C) $\int_0^\infty \dfrac{1}{x^2+6x+9} dx$

(D) All of the above

(E) None of the above

22. Let f be a continuous function on the interval $[-2, 4]$. If $f(-2) = 3$ and $f(4) = -3$, then the Mean Value Theorem guarantees that

(A) $f(0) = 0$
(B) $f'(c) = -1$ for some c between -2 and 4
(C) $f'(c) = 1$ for some c between -3 and 3
(D) $f(c) = 1$ for some c between -2 and 4
(E) $f(c) = 1$ for some c between -3 and 3

23. What is $\int \dfrac{5x+4}{2x^2+3x+1} dx$?

(A) $\ln|x+1| + \dfrac{3}{2} \ln|2x+1| + C$

(B) $2\ln|5x+4| + \dfrac{3}{2} \ln|2x+1| + C$

(C) $\ln|x+1| + 3\ln|2x+1| + C$

(D) $\ln|x+1| + 3\ln|5x+4| + C$

(E) $2\ln|x+1| + 3\ln|5x+4| + C$

Section II

Time	Number of Questions
45 Minutes	3

A GRAPHING CALCULATOR IS REQUIRED FOR SOME QUESTIONS ON THIS PART OF THE EXAMINATION

Directions: Solve each of the following problems using the available space for scratchwork. After examining the form of the choices, decide which is the best of the choices given and fill in the corresponding oval on the answer sheet. No credit will be given for anything written in the test book. Do not spend too much time on any one problem.

In this test: SHOW ALL YOUR WORK. You will be graded on the correctness and completeness of your methods as well as the accuracy of your final answers. Correct answers without supporting work may not receive credit.

You are permitted to use your calculator to solve an equation, find the derivative of a function at a point, or calculate the value of a definite integral. However, you must clearly indicate the setup of your program, namely the equation, function, or integral you are using. If you use other built-in features or programs, you must show the mathematical steps necessary to produce your results.

Unless otherwise specified, answers (numeric or algebraic) need not be simplified. If your answer is given as a decimal approximation, it should be correct to three places after the decimal point.

---

1. Consider a particle moving in the plane with $\frac{dx}{dt} = \sqrt{t+1}$ and $\frac{dy}{dt} = e^{-t}$, for $t \geq 0$.

   (A) What is the slope of the curve traced out by the particle, in terms of t?
   (B) If the particle is at the point (2, 3) when $t = 0$, where is it when $t = 3$? What is its speed then?
   (C) When is the speed of the particle the smallest?

2. Let $f(x)$ be a function that has derivatives of all orders for all real numbers. Assume $f(0) = 2$, $f'(0) = -4$, $f''(0) = 3$, and $f'''(0) = 5$.

   (A) Write the third-degree Taylor polynomial for $f(x)$ about $x = 0$.
   (B) Write the fifth-degree Taylor polynomial for $g(x)$, where $g(x) = xf(x^2)$ about $x = 0$, and use it to approximate $g(-0.2)$.
   (C) Write the third-degree Taylor polynomial for $h(x)$, where $h(x) = \int_0^x f(t)dt$, about $x = 0$.

3. The sides of a square are growing at a rate of e^{-t} cm per second, where t is measured in seconds; they are initially 1 cm long.

   (A) How long are the sides after t seconds?
   (B) How fast is the area of the square growing when $t = 3$?
   (C) What is the limit of the area of the square, as t approaches infinity?
   (D) What is the average side length during the first 5 seconds?

ADVANCED PLACEMENT CALCULUS AB

DIAGNOSTIC TEST ANSWERS AND EXPLANATIONS

Section I, Part A

1. **The correct answer is (A).** The antiderivative of x^n is $\dfrac{x^{n+1}}{n+1}$, so the antiderivative of $x^3 - 3x^2 + 4x - 1$ is $\dfrac{x^4}{4} - x^3 + 2x^2 - x + (C)$.

2. **The correct answer is (B).** The derivative of $f(x)$ is $3x^2 + 6x - 24$, which is 0 when $x = -4$ and $x = 2$. It is negative between those two values, and positive elsewhere.

3. **The correct answer is (A).** The substitution $u = 3x$ turns the problem into $\int_0^{\frac{\pi}{2}} \dfrac{1}{3} \sin u \, du$, and the antiderivative of $\sin u$ is $-\cos u$. As $\cos\left(\dfrac{\pi}{2}\right) = 0$ and $\cos 0 = 1$, the answer is $\dfrac{1}{3}$.

4. **The correct answer is (E).** The chain rule tells us that
$\dfrac{dy}{dx} = \dfrac{dy}{du} \times \dfrac{du}{dx} \times \dfrac{dy}{du} = 2u$, and $\dfrac{du}{dx} = 5\cos 5x$.

5. **The correct answer is (C).** The derivative $f'(x)$ is $1 - \dfrac{2}{x}$, which is 0 when $x = 2$. As it is positive when $x > 2$, and negative when $x < 2$, x is a local minimum; it must be the absolute minimum, as it is the only critical point. Plugging it in to $f(x)$ gives the desired value.

6. **The correct answer is (B).** Use the limit definition of the derivative to see that this is the derivative of $\ln x$ at $x = 1$.

7. **The correct answer is (B).** Since $\lim_{x \to 0} \frac{\sin 2x}{x} = 2\lim_{x \to 0} \frac{\sin 2x}{2x} = 2$ (I) is defined. Also, $x^2 + x - 6 = (x-2)(x+3)$, so

$$\lim_{x \to 2} \frac{x^2 + x - 6}{x - 2} = \lim_{x \to 2}(x+3) = 5$$ so (II) is defined. Limit (III) is not defined, as the denominator approaches 0 but the numerator does not.

8. **The correct answer is (E).** Use the quotient rule and the chain rule with $u = 3x$.

9. **The correct answer is (A).** Separating variables gives $\frac{dy}{y^2} = \frac{dx}{x}$. Now integrate both sides to get $-\frac{1}{y} = \ln|x| + c$. Plug in $x = 1$; $y = -1$ to determine that $c = 1$, and solve for y.

10. **The correct answer is (D).** This is a direct application of the mean value theorem.

11. **The correct answer is (E).** $\int_1^7 f(x)dx = \int_1^3 f(x)dx + \int_3^7 f(x)dx = 3$, and $\int_1^7 2f(x) - g(x)dx = 2\int_1^7 f(x)dx - \int_1^7 g(x)dx$

12. **The correct answer is (C).** $f(x)$ has critical points at $x = 2$ and $x = 4$. As f increases up to $x = 2$, decreases from $x = 2$ to $x = 4$, and increases after $x = 4$, only $x = 4$ is a minimum.

13. **The correct answer is (B).** $f(x)$ is concave up when $f''(x)$ is positive, which occurs when $f'(x)$ is increasing.

14. **The correct answer is (B).** The particle is stationary when the derivative, $\frac{2 - 2t^2}{(1+t^2)^2}$ is equal to 0.

Section I, Part B

15. **The correct answer is (D).** The best approximation for $f(1.2)$ is $f(1) + 0.2 \times f'(1)$.

16. **The correct answer is (B).** The average value is $\int_0^2 \dfrac{\frac{x}{x^2+1}\,dx}{2}$. This can be calculated directly using a direct-integral program on the calculator, as long as the error is small enough. Alternatively, the antiderivative of $\dfrac{x}{x^2+1}$ is $\dfrac{\ln(x^2+1)}{2}$.

17. **The correct answer is (C).** Using the product rule, $f'(x) = 3\sqrt{2x^2+1} + \dfrac{12x^2}{2\sqrt{2x^2+1}}$. Plug in $x = 2$ to find the answer.

18. **The correct answer is (C).** Label the distance from the first boat to the dock y, the distance from the second boat to the dock x, and the distance between the two boats s. Then $s^2 = x^2 + y^2$. Taking derivatives with respect to time gives $\dfrac{2s\,ds}{dt} = \dfrac{2x\,dx}{dt} + \dfrac{2y\,dy}{dt}$. When $t = 2$; $x = 6$; $y = 14$; $\dfrac{dx}{dt} = 3$, and $\dfrac{dy}{dt} = 7$. Calculate s and then $\dfrac{ds}{dt}$.

19. **The correct answer is (A).** Small rectangles, with base dx and height $\sqrt{x^3+1}$, become disks of volume $\pi(x^3 + 1)\,dx$ when rotated around the x-axis.

20. **The correct answer is (B).** At time t, the population of the colony is $100e^{t \ln 2/2}$; plug in $t = 3$.

21. **The correct answer is (C).** Inspect the graph of the function.

22. **The correct answer is (C).** Dividing top and bottom by x^2 gives $\lim\limits_{x \to \infty} \dfrac{3 + \frac{2}{x} + \frac{2}{x^2}}{4 + \frac{1}{x} + \frac{5}{x^2}}$, which is equal to $\dfrac{3}{4}$.

23. **The correct answer is (B).** The function x^2y is equal to $2x^4 - x^2$, which has derivative $8x^3 - 2x$. The derivative is 0 when $x = -\frac{1}{2}$, 0, and $\frac{1}{2}$; inspecting the sign of the derivative shows that $x = -\frac{1}{2}$ and $x = \frac{1}{2}$ are both local minima. The value at both points is $-\frac{1}{8}$.

Section II

1. **(A)** Exponential functions grow much faster than polynomial functions, so as x gets large, e^{x^2} gets much larger than $x - 1$. Thus $\lim_{x \to \infty} f(x) = 0$.

   (B) An absolute maximum must occur when $f'(x) = 0$. The derivative of $f(x)$ is $\dfrac{1 + 2x - 2x^2}{e^{x^2}}$, which is 0 when $x = \dfrac{(1 \pm \sqrt{3})}{2}$. As $f'(x)$ is negative when $x < \dfrac{(1-\sqrt{3})}{2}$ and when $x > \dfrac{(1+\sqrt{3})}{2}$, and is positive in between these values, $f(x)$ has a maximum when

   $$x = \frac{(1+\sqrt{3})}{2} = 1.366.$$

   (C) $g'(x) = f(x)$, so $g(x)$ has critical points when $f(x) = 0$. This only happens when $x = 1$. As $f(x)$ is negative when $x < 1$, and positive when $x > 1$, $x = 1$ is an absolute minimum for $g(x)$.

2. **(A)** The acceleration is positive when the velocity is increasing. This occurs when $0 < t < 0.75$ and when $1.5 < t < 2$.

   (B) When $t = 1$, the acceleration of the particle is approximated by a difference quotient, $\dfrac{v(1.25) - v(1)}{0.25}$, which is equal to -8.

   (C) The average velocity of the particle is equal to $\int_0^2 \dfrac{v(t)\,dt}{2}$. We do not have enough information to calculate this precisely, but can do so using a Riemann sum with 8 rectangles, with the height of each rectangle given by the left endpoint:

   $$\frac{0.25(v(0) + v(0.25) + v(0.5) + v(0.75) + v(1) + v(1.25) + v(1.5) + v(1.75))}{2}$$

   which is equal to $\dfrac{9}{4}$. The average acceleration can be determined precisely; it is $\dfrac{v(2) - v(0)}{2 - 0} = \dfrac{5 - 1}{2} = 2$.

3. Graphing the two functions shows that they cross twice, and that between those crossing points, $y = -x^2 + 4x - 3$ is on top. They cross when $-x^2 + 4x - 3 = x - 1$, which occurs when $x = 1$ and when $x = 2$.

(A) The distance from the top to the bottom is given by $-x^2 + 4x - 3 - (x - 1) = -x^2 + 3x - 2$. The derivative of this function is $-2x + 3$, which is equal to 0 when $x = \frac{3}{2}$. As the derivative is positive when $x < \frac{3}{2}$ and negative when $x > \frac{3}{2}$, there is an absolute maximum at $x = \frac{3}{2}$.

(B) The area between the curves is given by

$$\int_1^2 -x^2 + 4x - 3 - (x-1)dx = \int_1^2 -x^2 + 3x - 2\, dx$$

$$= \left.\frac{-x^3}{3} + \frac{3x^2}{2} - 2x\right|_1^2$$

$$= \frac{1}{6}$$

(C) By the method of shells, the volume of the solid of revolution is given by

$$2\pi\int_1^2 x\left((-x^2 + 4x - 3) - (x - 1)\right)dx = 2\pi\int_1^2 -x^3 + 3x^2 - 2x\, dx.$$

This integral is equal to $2\pi\left(-x^4/4 + x^3 - x^2\right)\Big|_1^2 = \frac{\pi}{2}$.

ADVANCED PLACEMENT CALCULUS BC

DIAGNOSTIC TEST ANSWERS AND EXPLANATIONS

Section I, Part A

1. **The correct answer is (A).** By the chain rule, $\dfrac{dy}{dx} = \dfrac{dy}{dt} \times \dfrac{dt}{dx}$.

    Calculate derivatives; $\dfrac{dy}{dt} = \dfrac{-1}{(1-t)}$, and $\dfrac{dx}{dt} = 3t^2$, so $\dfrac{dt}{dx} = \dfrac{1}{(3t^2)}$.

2. **The correct answer is (B).** The derivative of $f(x)$ is $\dfrac{3\sqrt{x+1}}{2} - 2xe^{x^2-9}$; plug in $x = 3$ to get -3.

3. **The correct answer is (E).** The slope of the tangent line is the derivative of the curve, which is equal to $\dfrac{2x}{(x^2+1)}$. Plug in $x = 3$ to get the answer.

4. **The correct answer is (D).** Use integration by parts, with $u = \ln x$ and $dv = \dfrac{1}{\sqrt{x}}$, to turn this integral into $2\sqrt{x}\ln x - \int \dfrac{2\sqrt{x}}{x}\,dx$.

5. **The correct answer is (C).** Divide top and bottom by x^2 to get

    $$\dfrac{3\sqrt{1-\dfrac{3}{x^3}}}{2+\dfrac{(\cos x)}{x^2}}$$

    everything except the constants goes to 0.

6. **The correct answer is (A).** The Taylor series about 0 of $\dfrac{1}{(1-x)}$ is $1 + x + x^2 + x^3 + \ldots$ Integrate this to get the series for $-\ln(1-x)$, multiply by -1 and replace x with $-x^2$ to get the answer.

7. **The correct answer is (D).** As water is pouring out of the funnel, the height of the water is decreasing, hence $h'(t)$ is negative. As the water pours out at a constant rate, the height is decreasing more quickly over time, so $h''(t)$ is also negative. However, the height certainly is positive.

8. **The correct answer is (E).** Make the substitution $u = x^2 - 1$ to turn the integral into $\int_0^8 \frac{\sqrt{u}}{2} du$, which is equal to $\frac{1}{3} u^{3/2} \Big|_0^8$.

9. **The correct answer is (E).** For $f(x)$ to be continuous at 3, $f(3)$ must be equal to $\lim_{x \to 3} f(x)$. Factoring $x - 3$ out of the numerator of $f(x)$ shows that this limit is equal to 0.

10. **The correct answer is (C).** (I) guarantees that the sum diverges, by the ratio test. (II) guarantees that it converges, by the integral test. (III) guarantees that it diverges by the comparison test, as $\sum_{n=1}^{\infty} \frac{1}{n}$ diverges.

11. **The correct answer is (D).** The function has a vertical tangent line when the derivative has a vertical asymptote. Taking derivatives gives $\frac{dy}{dx} = -\frac{5x+y}{x+y}$. This has a vertical asymptote when $x = -y$. Plugging this back in to the original equation, we get $5x^2 - 2x^2 + x^2 = 16$, which happens when $x = \pm 2$.

12. **The correct answer is (B).** An estimate of $f(0.95)$ is given by $f(1) - 0.05 f'(1)$.

13. **The correct answer is (A).** The function has a point of inflection where the derivative has a critical point. The derivative must have a local minimum between $x = 1$ and $x = 3$, as it must decrease at some point between $x = 1$ and $x = 2$, and must increase at some point between $x = 2$ and $x = 3$, so (I) must be true. However, it could decrease steadily from $x = 0$ to $x = 2$, and increase steadily after that, so neither (II) nor (III) need be true.

14. **The correct answer is (E).** By comparison with a p-series, (A) and (B) converge. By the alternating series test, (C) converges.

Section I, Part B

15. **The correct answer is (A).** Plug $\frac{dx}{dt} = e^t$ and $\frac{dy}{dt} = 1$ into the formula for arc length to get $\int_0^3 \sqrt{e^{2t}+1}\,dt$, and evaluate that on the calculator.

16. **The correct answer is (C).** The series converges when $|x - 5| < 3$.

17. **The correct answer is (C).** The second derivative of the position function is $-9\cos(3t) + \frac{4}{t^2}$; graph this on the calculator and count the zeros between $t = 0$ and $t = 3$.

18. **The correct answer is (B).** After t days, there are $5e^{kt}$ grams left, for some k. We know that $5e^{4k} = 2.5$, so $k = -0.173$. Now solve $5e^{-0.173t} = 1$ for t.

19. **The correct answer is (D).** The curve has radius 0 when $\theta = 0$ and when $\theta = \pi$, so the area is given by $\int_0^\pi \frac{(3\sin\theta)^2}{2} \cdot d\theta$. Evaluate this on the calculator.

20. **The correct answer is (A).** Graph $f(x)$. It is always positive, so the area under it from 1 to x increases at x increases. Thus (I) is true. The graph clearly shows that $f(x)$ decreases when x is positive, so (II) is false. Similarly, the slope goes from positive to negative, so (III) is false.

21. **The correct answer is (E).**

$\int_1^3 (2x-2)^{-1/3}\,dx = \lim_{x \to 1} \frac{3}{4}(2x-2)^{2/3} - \frac{3}{4}4^{2/3}$, which converges.

$\int_{-\infty}^3 x^2 e^{x^3}\,dx = \frac{1}{3}\int_{-\infty}^{27} e^u\,du$, which is equal to $\lim_{x \to -\infty} \frac{1}{3}\left[e^u - e^{27}\right]$, which converges. As $x^2 + 6x + 9$ is degree bigger than 1, (C) converges. Thus none of them diverge.

22. **The correct answer is (B).** The Mean Value Theorem guarantees that $f'(c) = \frac{(f(4) - f(-2))}{6}$ for some c between -2 and 4.

23. **The correct answer is (A).** The method of partial fractions turns this into $\int \frac{1}{x+1} + \frac{3}{2x+1}\,dx$

Section II

1. (A) The slope of the curve is $\dfrac{dy}{dx} = \dfrac{dy}{dt}\dfrac{dt}{dx}$ so $\dfrac{dy}{dx} = \dfrac{e^{-t}}{\sqrt{t+1}}$.

   (B) The position of the particle is given by the anti-derivative of the velocity. Thus $x(t) = \dfrac{2}{3}(t+1)^{3/2} + C$. As $x(0) = 2$, $C = \dfrac{4}{3}$. Similarly, $y(t) = -e^{-t} + (C)$. As $y(0) = 3$, $c = 4$. Thus, when $t = 3$, the particle is at position $\left(\dfrac{20}{3}, 4 - e^{-3}\right)$. Its speed is $\sqrt{\left(\dfrac{dx}{dt}\right)^2 + \left(\dfrac{dy}{dt}\right)^2}$, so when $t = 3$, the speed is $\sqrt{4 + e^{-6}}$

   (C) The speed of the particle, $s(t)$, is equal to $\sqrt{t + 1 + e^{-2t}}$. It has a critical point when $s'(t) = 0$. The derivative, $s'(t)$, is $\dfrac{1 - 2e^{-2t}}{2\sqrt{t+1+e^{-2t}}}$, which is equal to 0 when $t = \dfrac{(\ln 2)}{2}$. As $s'(t)$ is negative when t is less than $\dfrac{(\ln 2)}{2}$, and positive when t is bigger than $\dfrac{(\ln 2)}{2}$, this is a minimum.

2. (A) The third-degree Taylor polynomial for $f(x)$ is $2 - 4x + \dfrac{3}{2}x^2 - \dfrac{5}{6}x^3$.

   (B) The fifth-degree Taylor polynomial for $g(x)$ is $2x - 4x^3 + \dfrac{3}{2}x^5$. Plug in $x = 0.2$ to get $g(0.2) \approx 0{:}368$.

   (C) The third-degree Taylor polynomial for $h(x)$ is $2x - 2x^2 + \dfrac{1}{2}x^3$.

3. (A) The length of the sides is the anti-derivative of the rate of growth, so they are $s(t) = -e^{-t} + c$ centimeters long. As $s(0) = 1$, $c = 2$, so $s(t) = 2 - e^{-t}$.

(B) The area of the square, $a(t)$ is equal to $s(t)^2$, so $\dfrac{da}{dt} = 25(t) \cdot \dfrac{ds}{dt}$. When $t = 3$, this is equal to $2(2 - e^{-3})e^{-3}$.

(C) The limit of the area is $\lim\limits_{t \to \infty}(2 - e^{-t})^2 = 4$, as $\lim\limits_{t \to \infty} e^{-t} = 0$

(D) The average side length during the first five seconds is

$$\frac{1}{5}\int_0^5 2 - e^{-t}\, dt = \frac{1}{5}\left(9 + e^{-5}\right)$$

Part I
FUNCTIONS, GRAPHS, AND LIMITS

FUNCTIONS

A **function** is a set of ordered pairs where each first coordinate is paired with a unique second coordinate. If every vertical line passes through the graph of an equation at most once, that equation represents a function (this is called the vertical line test). The set of all first coordinates of a function is called the **domain**, the set of second coordinates is called the **range**. Equations that represent functions are generally written using function notation: $f(x) = 2x + 1$ or $g(t) = 4t^2 - 1$. In these examples, x and t are independent variables, and $f(x)$ and $g(t)$ represent dependent variables.

Example 1
Determine which of the following are functions. For those that are functions, find the domain and range, stating the answers in interval notation.

a.

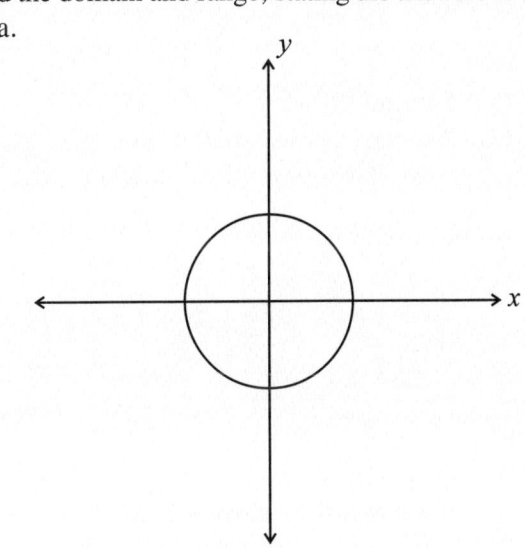

b.

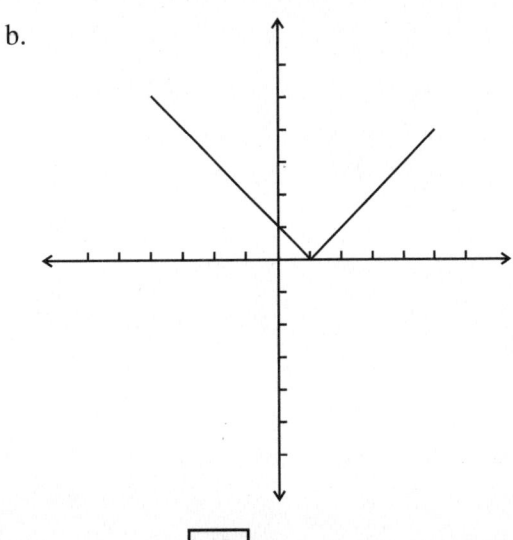

c. $3x + 2y = 6$

Solutions:

a. This is not a function since a vertical line intersects the graph at more than one point.

b. The graph represents a function. The domain is $(-\infty, \infty)$. Note that the screen shows only a portion of the graph, and it is assumed the graph continues in a similar fashion. The range is $[0, \infty)$, found by noting that the smallest y-value shown is 0, and the graph appears to rise without bound.

c. This equation is of the form $ax + by = c$ and represents a linear function. It can be solved for y as $y = -\frac{3}{2}x + 3$ and then graphed on your graphing calculator. The domain is $(-\infty, \infty)$, and the range is $(-\infty, \infty)$.

Example 2
Assume a particle moves along the x-axis so that its position at any time $t \geq 0$ is given by $g(t) = 4t^2 + 1$. Find the position of the particle at $t = 0$ and $t = 3$.

Solution:

When $t = 0$, $g(0) = 4(0)^2 + 1 = 1$.
When $t = 3$, $g(3) = 4(3)^2 + 1 = 36 + 1 = 37$.

TRANSLATION AND SYMMETRY

Graphs of several basic functions are shown below using a standard viewing window $-10 \leq x \leq 10$, $-10 \leq y \leq 10$, Xscl = 1, Yscl = 1 (see your calculator manual if you need help setting the viewing window). The graph of $y = f(x)$ can be transformed as follows for $a > 0$.

$f(x) + a$ shifts $f(x)$ upward a units
$f(x) - a$ shifts $f(x)$ downward a units
$f(x - a)$ shifts $f(x)$ right a units
$f(x + a)$ shifts $f(x)$ left a units
$-f(x)$ reflects $f(x)$ about the x axis
$f(-x)$ reflects $f(x)$ about the y axis

FUNCTIONS, GRAPHS, AND LIMITS

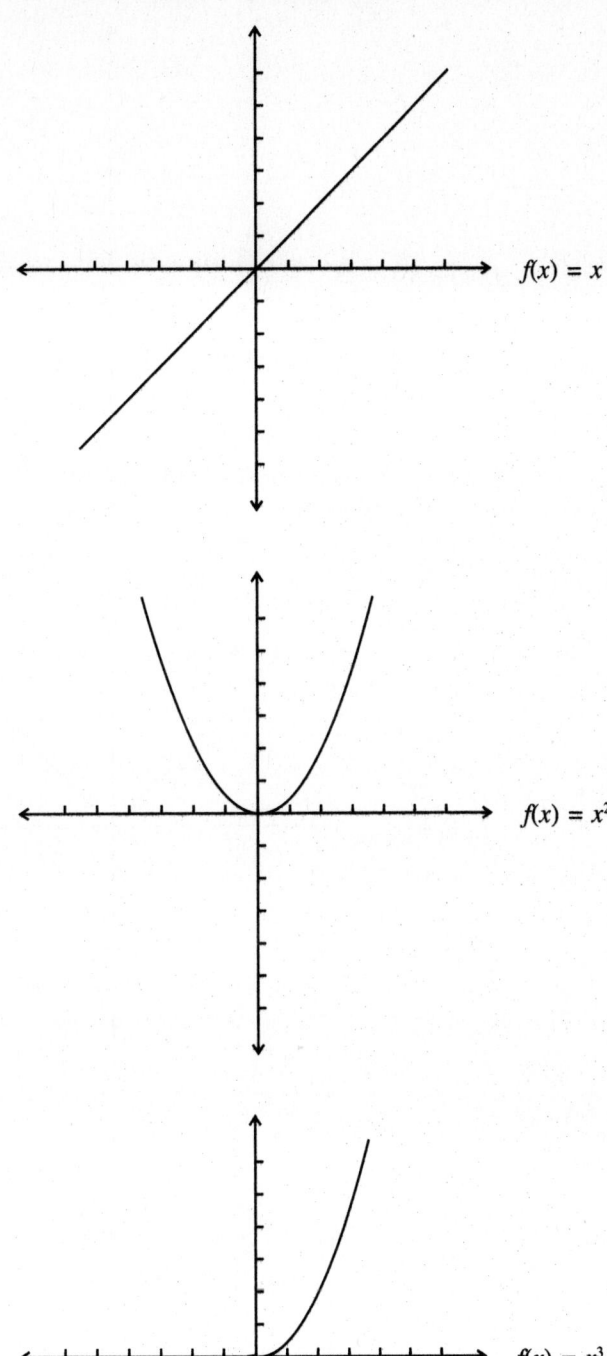

PART I

$f(x) = \sqrt{x}$

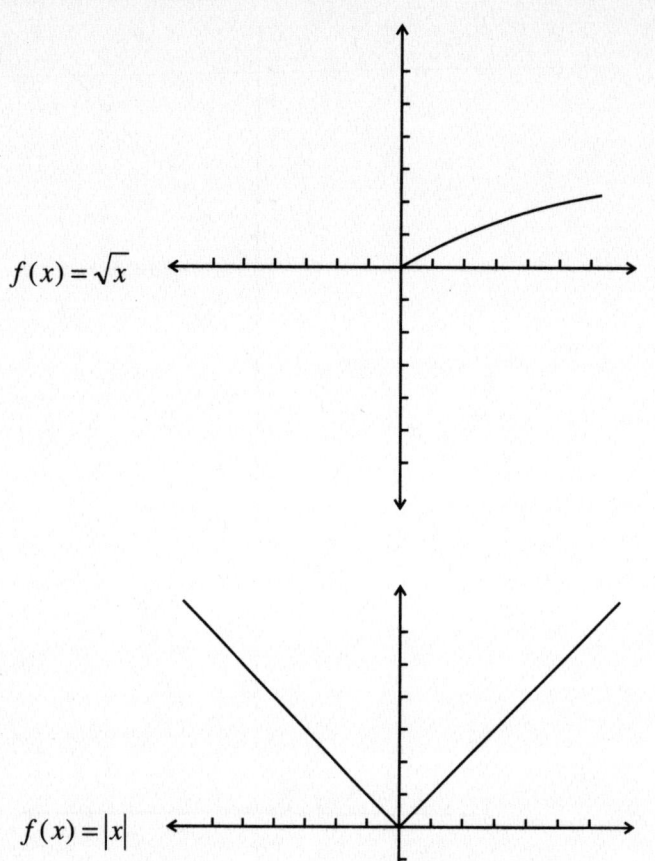

$f(x) = |x|$

$f(x) = \dfrac{1}{x}$

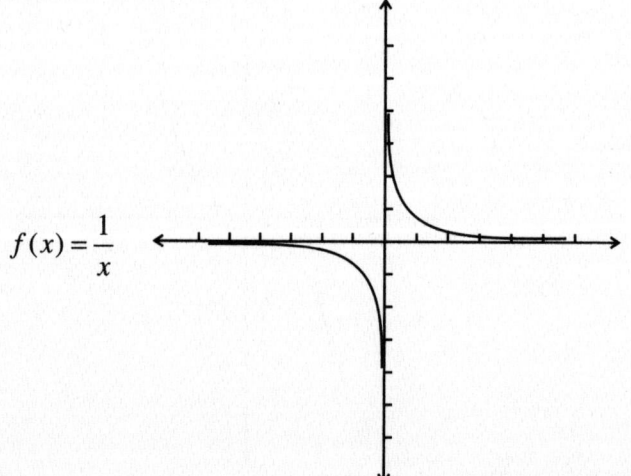

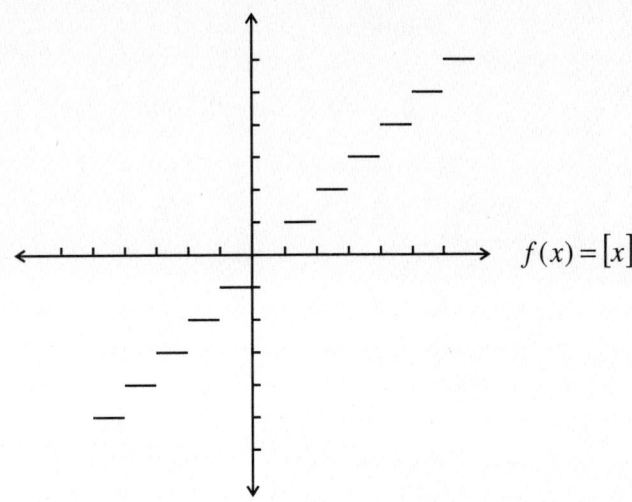

Graphs are symmetric with respect to the y axis if $f(-x) = f(x)$ and symmetric with respect to the origin if $f(-x) = -f(x)$. A function that is symmetric with respect to the y axis is called an even function. A function that is symmetric with respect to the origin is called an odd function. Note that $f(x) = x^2$ and $f(x) = |x|$ are even functions, while $f(x) = x, f(x) = x^3$, and $f(x) = \dfrac{1}{x}$ are odd functions. The functions $f(x) = \sqrt{x}$ and $f(x) = [x]$ are neither even nor odd.

Example 3
Use the graph of $y = f(x)$ shown below to sketch the graph of
 a. $y = f(x) + 2$
 b. $y = f(x + 2)$
 c. $y = |f(x)|$
 d. $y = -f(x)$
 e. $y = f(-x)$

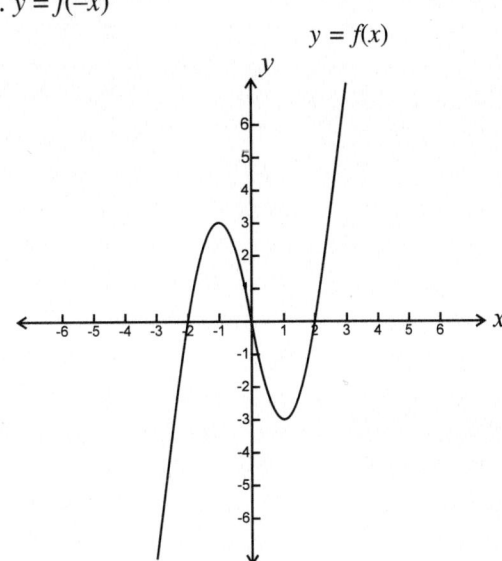

PART I

Solutions:

a. $y = f(x) + 2$ will shift the graph upward 2 units:

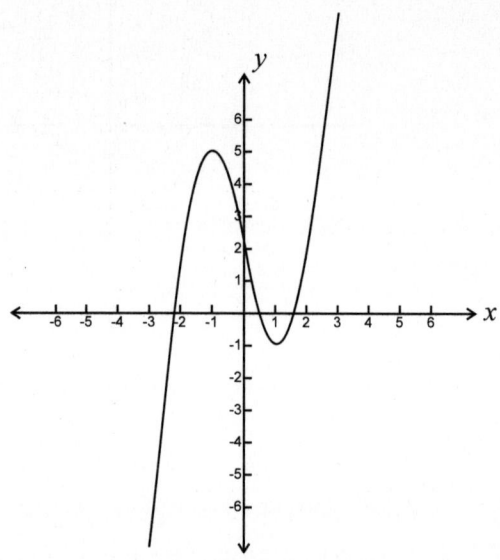

b. $y = f(x + 2)$ will shift the graph to the left 2 units:

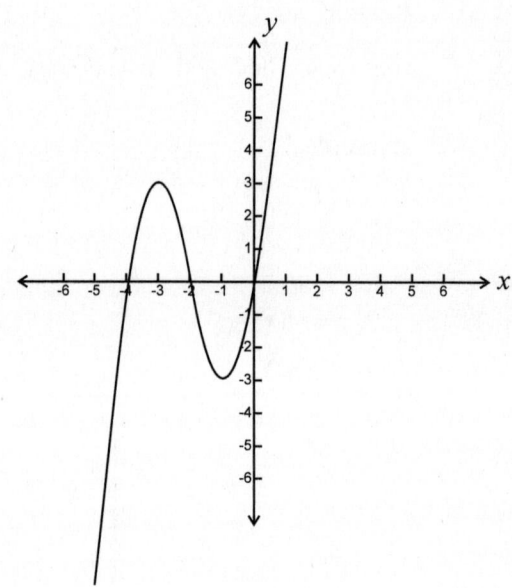

c. $y = |f(x)|$ will make the negative y values into positive y values:

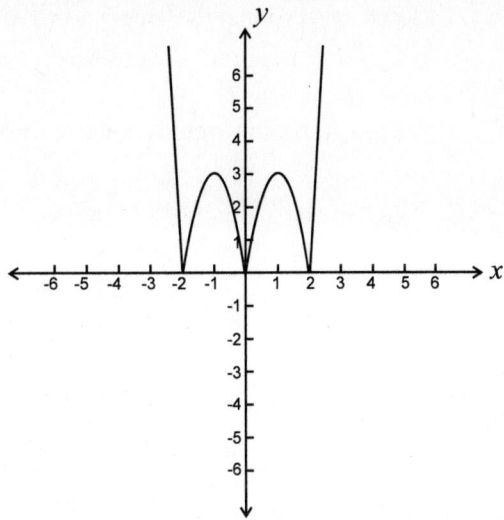

d. $y = -f(x)$ will reflect the graph about the x axis:

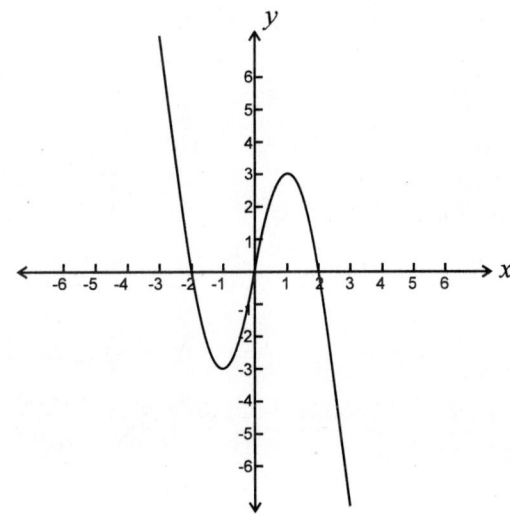

e. $y = f(-x)$ will reflect the graph about the y axis:

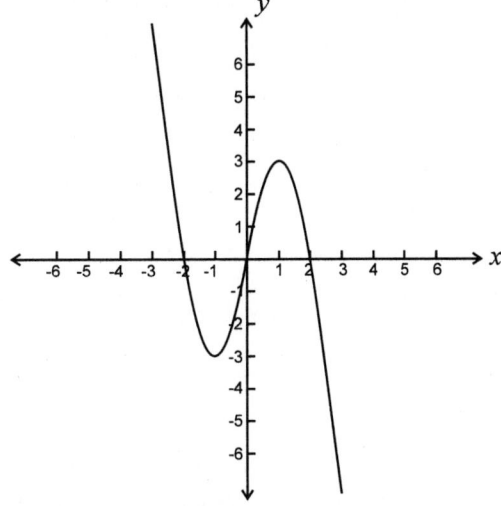

Note that the given function appears to be an odd function and hence $f(-x) = -f(x)$ so that the graphs of (d) and (e) are the same.

Example 4
Use the graph shown to write an expression for $f(x)$ in terms of x.

a.
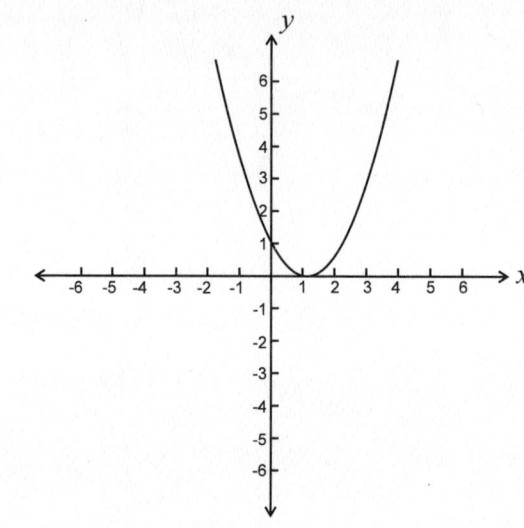

b.
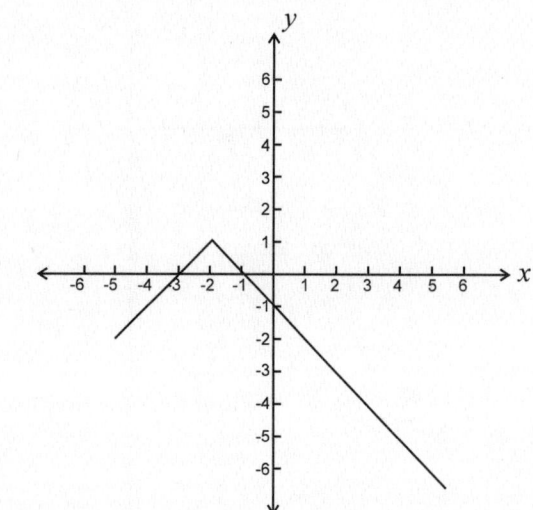

Solutions:

a. The graph appears to be a parabola with a vertex at (1, 0). Using shifting techniques, the vertex has been shifted one unit to the right, and thus we try $f(x) = (x-1)^2$ as the equation. We also see that (0, 1) is a point on the graph, and check (0, 1) in our proposed equation: $1 = (0-1)^2$ and $1 = 1$.

FUNCTIONS, GRAPHS, AND LIMITS

b. The graph appears to be an absolute value graph with vertex $(-2, 1)$. The basic form of $f(x) = |x|$ must show the shift of the vertex 2 units to the left, upward 1 unit, as well as the reflection across the x axis: $f(x) = -|x + 2| + 1$. Check the point $(-1, 0)$: $0 = -|-1 + 2| = -1 + 1$.

Example 5

Determine whether the given function is even, odd, or neither.

a. $f(x) = x^4 + 2x^2$
b. $g(x) = 4x^3 - x$
c. $h(x) = \begin{cases} x^4 & \text{for } x \leq 0 \\ x^2 & \text{for } x > 0 \end{cases}$

Solutions:

a. Since $f(-x) = (-x)^4 + 2(-x)^2 = x^4 + 2x^2$, $f(-x) = f(x)$, f is an even function. Or, by graphing, observe that the graph is symmetric with respect to the y axis and hence is an even function. Note that all the exponents are even.

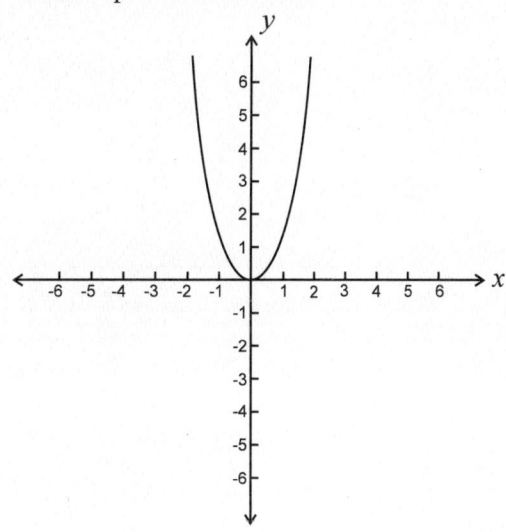

b. $g(-x) = 4(-x)^3 - (-x) = -4x^3 + x = -(4x^3 - x)$. Thus $g(-x) = -g(x)$, which means g is an odd function. Or, by graphing, observe that the graph is symmetric with respect to the origin and hence is an odd function. Note that all the exponents are odd.

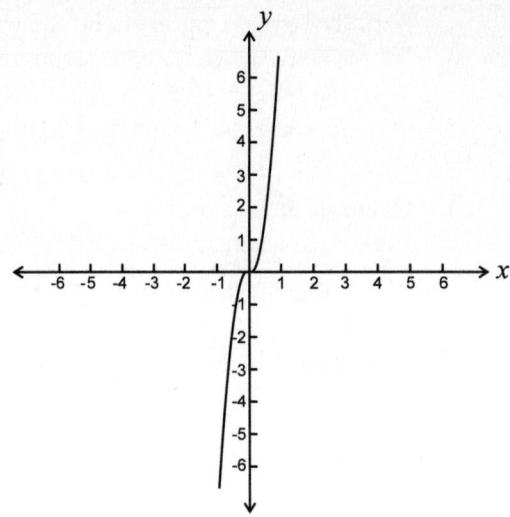

c. $h(-x) = \begin{cases} x^2 & \text{for } x \leq 0 \\ x^4 & \text{for } x > 0 \end{cases}$ which means $h(-x) \neq -h(x)$. Thus h is neither even nor odd. The graph approach is probably easier for h, since the graph is not symmetric with respect to either the y axis or the origin.

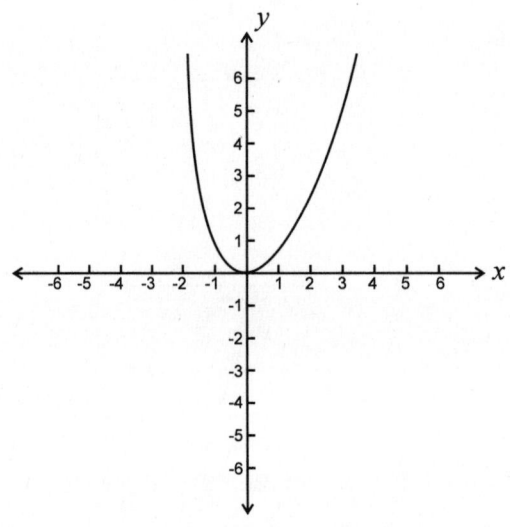

OPERATIONS WITH FUNCTIONS

Functions can be combined using the usual operations of addition, subtraction, multiplication, and division, as well as by composition as defined below.

If f and g are functions,

$$(f + g)(x) = f(x) + g(x)$$
$$(f - g)(x) = f(x) - g(x)$$

$$\left(\frac{f}{g}\right)(x) = \frac{f(x)}{g(x)}, \quad g(x) \neq 0$$

FUNCTIONS, GRAPHS, AND LIMITS

Example 6
If $f(x) = 4x^2 + 1$, $g(x) = 4x$, $h(x) = |x|$

a. Describe the graph of $r(x) = (f+g)(x)$

b. Find $(g-h)(-3)$

c. Find the domain of $\left(\dfrac{f}{g}\right)(x)$

d. Find $(f \circ g)(x)$

e. Find $(f \circ h)(x)$

Solutions:

a. Since $(f+g)(x) = f(x) + g(x) = (4x^2 + 1) + 4x$, $r(x) = 4x^2 + 4x + 1$. A quick check with a graphing calculator reveals that the graph is a parabola, with vertex $\left(-\dfrac{1}{2}, 0\right)$, opening upward, with domain $(-\infty, \infty)$.

b. $(g-h)(-3) = g(-3) - h(-3) = -12 - (|-3|) = -12 - 3 = -15$

c. Since $\left(\dfrac{f}{g}\right)(x) = \dfrac{4x^2 + 1}{4x}$, the domain excludes values where the denominator equals 0. $4x = 0$ when $x = 0$, so the domain is $(-\infty, 0) \cup (0, \infty)$.

d. $(f \circ g)(x) = f(g(x)) = f(4x) = 4(4x)^2 + 1 = 64x^2 + 1$

e. $(f \circ h)(x) = f(h(x)) = f(|x|) = 4(|x|)^2 + 1 = 4x^2 + 1$ since $|x|^2 = x^2$

POLYNOMIAL FUNCTIONS

A polynomial function can be written as $f(x) = a_n x^n + a_{n-1} x^{n-1} + \ldots + a_2 x^2 + a_1 x + a_0$, $a_n \neq 0$. The leading coefficient, a_n, and the degree, n, determine the left and right behavior of the graph. The following possibilities exist:

	Degree	Leading Coefficient	$x \to -\infty$	$x \to \infty$	Thinking of the graph of
1.	even	+	$f(x) \to +\infty$	$f(x) \to +\infty$	$f(x) = x^2$
2.	even	−	$f(x) \to -\infty$	$f(x) \to -\infty$	$f(x) = -x^2$
3.	odd	+	$f(x) \to -\infty$	$f(x) \to +\infty$	$f(x) = x^3$
4.	odd	−	$f(x) \to +\infty$	$f(x) \to -\infty$	$f(x) = -x^3$

Rational Functions

A rational function can be written as $f(x) = \dfrac{p(x)}{q(x)}$, $q(x) \neq 0$ where $p(x)$ and $q(x)$ are polynomial functions. While the graph of any polynomial function is continuous (no breaks), the graph of a rational function may have one or more asymptotes. Vertical asymptotes occur where the denominator of the rational function equals 0. (If the same value makes both the numerator and denominator equal 0, the graph contains a hole at that value rather than a vertical asymptote.) To find horizontal asymptotes for the graphs of rational functions without the use of calculus, compare the degree of the numerator to the degree of the denominator:

If $f(x) = \dfrac{a_m x^m + \ldots + a_0}{b_n x^n + \ldots + b_0}$ and

a. $m > n$, the graph contains no horizontal asymptote.

b. $m = n$, the graph contains a horizontal asymptote at $y = \dfrac{a_m}{b_n}$.

c. $m < n$, the graph contains a horizontal asymptote at $y = 0$.

The topic of asymptotes will be discussed again in Chapter 2.

Example 7
The graph of $f(x) = x^3 + 10x^2 - 24x$ is shown below in a standard viewing window. Use your knowledge of the behavior of a polynomial as x increases without bound to discuss the graph.

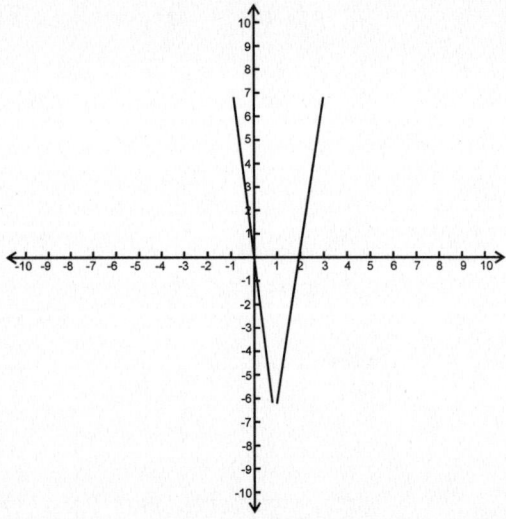

FUNCTIONS, GRAPHS, AND LIMITS

Solution:

The degree of $f(x) = x^3 + 10x^2 - 24x$ is odd (3) and the leading coefficient is positive (1). Thus as x approaches $-\infty$, $f(x)$ should approach $-\infty$. The standard viewing window is not showing a sufficient portion of the graph. Changing the range to $-20 \le x \le 20$ with a scale of 5, $100 \le y \le 400$ with a scale of 50 produces:

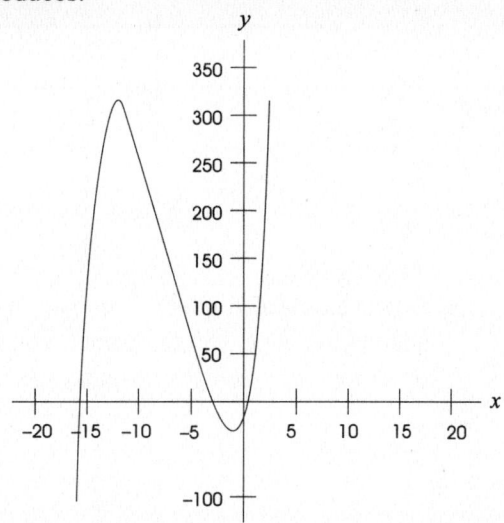

which then agrees with the appropriate left and right behavior for a polynomial of odd degree with a positive leading coefficient.

Example 8
Find the horizontal and vertical asymptotes for each function.

a. $f(x) = \dfrac{x^2 - 1}{x + 2}$

b. $g(x) = \dfrac{4x - 5}{2x - 3}$

c. $f(x) = \dfrac{2x + 5}{x^2 - 2x - 8}$

d. $p(x) = (x - 3)^2 - 1$

Solutions:

a. Since the degree of the numerator is greater than the degree of the denominator (2 > 1), there is no horizontal asymptote. Setting the denominator equal to 0 and solving gives $x = -2$ for the vertical asymptote. If your calculator is set in connect mode, it appears to show the vertical asymptote.

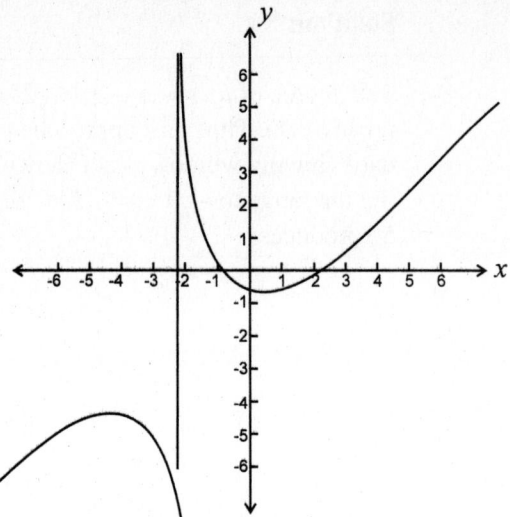

However, the asymptote is *not* part of the graph and this is your calculator's attempt to connect all the points. You may find it easier to use Dot mode (refer to your calculator manual) to view rational functions.

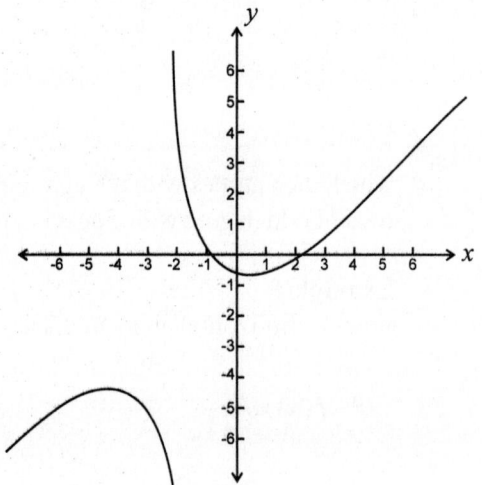

b. Since the degree of the numerator equals the degree of the denominator (1), there will be a horizontal asymptote at $y = \frac{4}{2} = 2$. Setting the denominator equal to 0 and solving gives $x = \frac{3}{2}$ for the vertical asymptote. Use your calculator to confirm this.

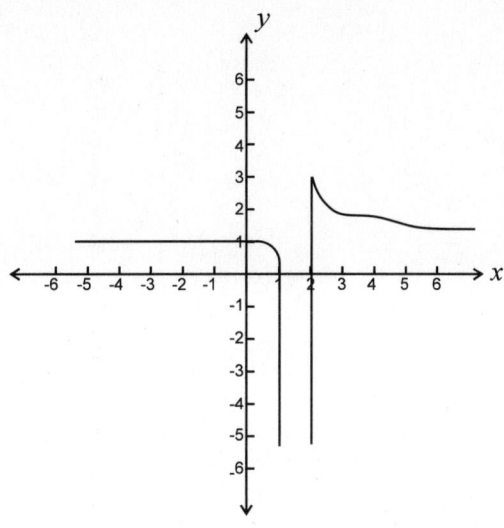

Here is the graph drawn with the asymptotes shown as dotted lines (not done on a graphing calculator).

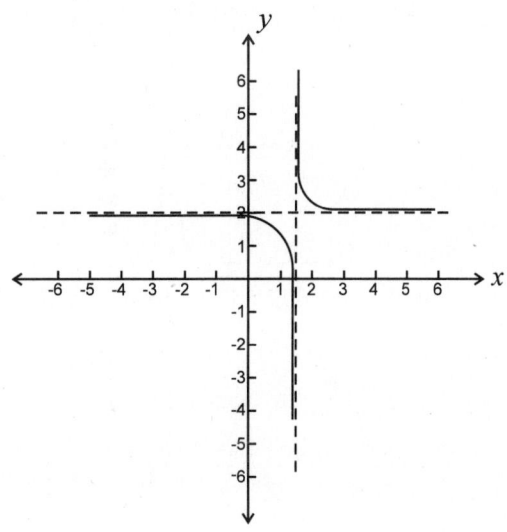

c. Since the degree of the numerator is less than the degree of the denominator (1 < 2), there is a horizontal asymptote at $y = 0$. Setting the denominator equal to 0 and solving (either by factoring or using the quadratic formula) gives $x = 4$ and $x = -2$ as the vertical asymptotes. Use your calculator to confirm this.

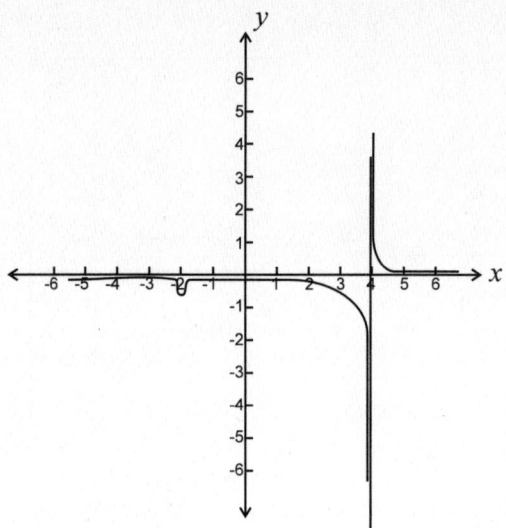

Here is the graph drawn with the asymptotes shown as dotted vertical lines.

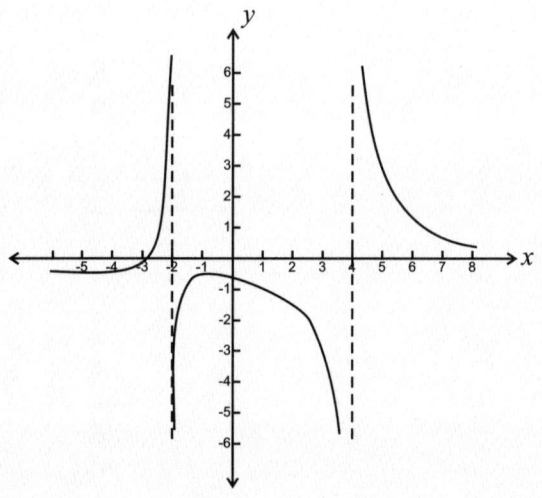

d. $p(x)$ is not a rational function. You should recognize this as a parabola, and hence there are no horizontal or vertical asymptotes. There are other functions whose graphs have asymptotes (including some trigonometric functions, exponential and logarithmic functions), but polynomial functions do not have asymptotes.

Example 9

If the graph of $y = \dfrac{ax+b}{x+c}$ has a horizontal asymptote $y = 3$ and a vertical asymptote $x = -4$, find a, b, and c.

FUNCTIONS, GRAPHS, AND LIMITS

Solution:

The vertical asymptote is found by setting the denominator equal to 0, so $x + c = 0$ when $x = -4$, which means $-4 + c = 0$ or $c = 4$. Since the degree of the numerator equals the degree of the denominator, the horizontal asymptote is $y = \frac{a}{1} = 3$ so $a = 3$. There is not enough information given to enable you to find b.

TRIGONOMETRIC FUNCTIONS

Many of the identities studied in a trigonometry course will be used in calculus. In particular, you should know:

$$\sin\theta = \frac{1}{\csc\theta} \qquad \sin^2\theta + \cos^2\theta = 1$$

$$\cos\theta = \frac{1}{\sec\theta} \qquad 1 + \tan^2\theta = \sec^2\theta$$

$$\tan\theta = \frac{1}{\cot\theta} = \frac{\sin\theta}{\cos\theta} \qquad 1 + \cot^2\theta = \csc^2\theta$$

$$\sin(\alpha \pm \beta) = \sin\alpha\cos\beta \pm \cos\alpha\sin\beta \qquad \sin 2\theta = 2\sin\theta\cos\theta$$

$$\cos(\alpha \pm \beta) = \cos\alpha\cos\beta \mp \sin\alpha\sin\beta \qquad \cos 2\theta = 2\cos^2\theta - 1 = 1 - 2\sin^2\theta$$

$$\tan(\alpha \pm \beta) = \frac{\tan\alpha \pm \tan\beta}{1 \mp \tan\alpha\tan\beta} \qquad = \cos^2\theta - \sin^2\theta$$

The graphs of the trigonometric functions are accessible from a graphing calculator—be sure your calculator is in radian mode (check the calculator manual if you need help). While $y = \sin x$, $y = \cos x$, and $y = \tan x$ are available with the SIN, COS, and TAN buttons, use the identities to graph the reciprocal functions (i.e., $y = \csc x = \frac{1}{\sin x}$).

A summary of other graphing information follows.

The graph of $y = a \sin(bx + c)$ has amplitude $|a|$, period $\frac{2\pi}{b}$, phase shift $-\frac{c}{b}$, and is reflected over the x axis if $a < 0$.

PART I

The graph of $y = a\cos(bx + c)$ has amplitude $|a|$, period $\frac{2\pi}{b}$, phase shift $-\frac{c}{b}$, and is reflected over the x axis if $a < 0$. The graph of $y = a\tan(bx + c)$ has period $\frac{\pi}{b}$ and phase shift $-\frac{c}{b}$.

Example 10

Find the amplitude and period for each function. Then sketch the graph of each function on the indicated interval, using dotted lines for any asymptotes.

a. $y = 3\sin(4x)$ on $[-2\pi, 2\pi]$

b. $y = -4\cos(2x)$ on $[-2\pi, 2\pi]$

c. $y = \sec x$ on $[-2\pi, 2\pi]$

d. $y = \tan(2x)$ on $[-\pi, \pi]$

Solutions:

a. amplitude $= 3$, period $= \frac{2\pi}{4} = \frac{\pi}{2}$

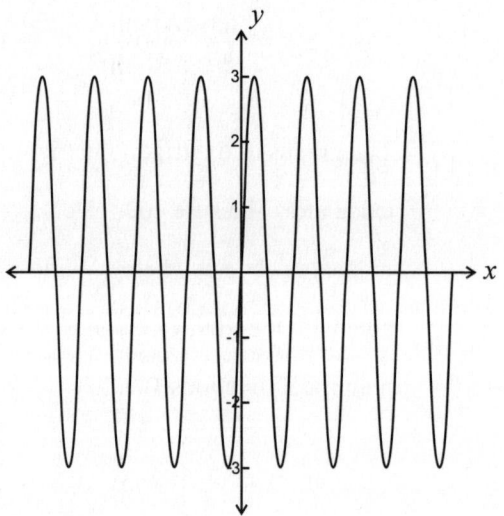

b. amplitude = |–4| = 4, period = $\dfrac{2\pi}{2} = \pi$

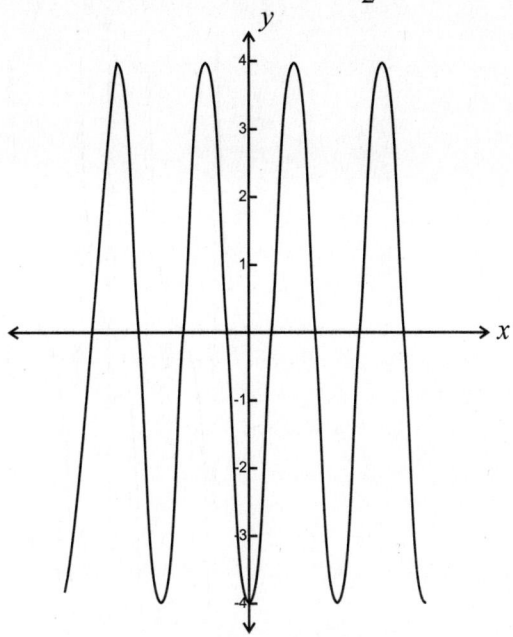

c. amplitude is not defined, period = 2π

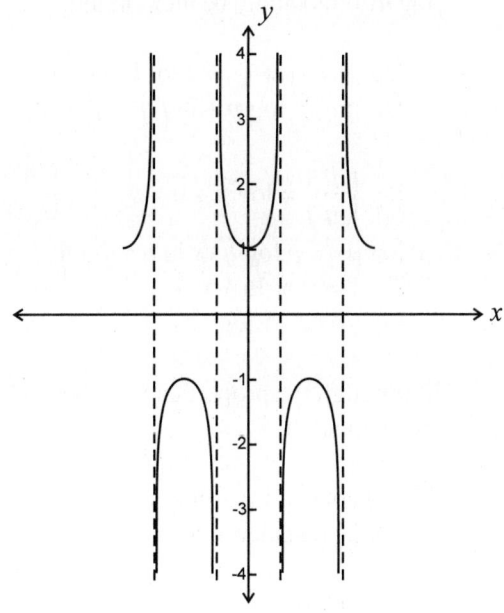

d. amplitude is not defined, period $= \dfrac{\pi}{2}$

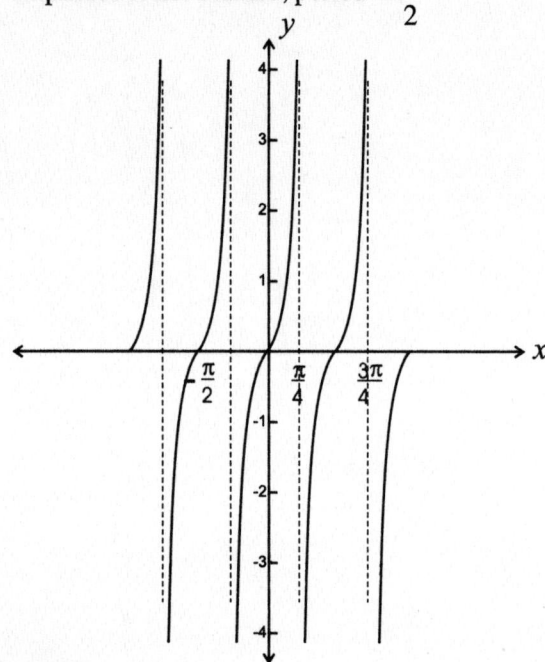

EXPONENTIAL AND LOGARITHMIC FUNCTIONS

The definition of a logarithm and basic properties used in working with logarithms should be memorized.

$\log_b x = y \Leftrightarrow b^y = x$ for $b > 0$, $b \neq 1$, $x > 0$

$\log(pq) = \log p + \log q$ or $\ln(pq) = \ln p + \ln q$

$\log\left(\dfrac{p}{q}\right) = \log p - \log q$ or $\ln\left(\dfrac{p}{q}\right) = \ln p - \ln q$

$\log a^p = p \log a$ or $\ln a^p = p \ln a$

$\log 1 = 0$ or $\ln 1 = 0$

$\log_a a = 1$ so $\log 10 = 1$ and $\ln e = 1$

(Although the properties have been stated for base 10 and base e, other bases can be used.)

The graphs of logarithmic and exponential functions are accessible from a graphing calculator. Some graphing features are summarized below.

The graph of $y = \log_a x$ has a vertical asymptote at $x = 0$, has domain $(0, \infty)$, range $(-\infty, \infty)$.

The graph of $y = a^x$ has a horizontal asymptote at $y = 0$, has domain $(-\infty, \infty)$, range $(0, \infty)$.

FUNCTIONS, GRAPHS, AND LIMITS

Example 11
Use the properties of logarithms to write each expression as a sum or difference of logarithms.

a. $\log 6x^3$

b. $\ln \dfrac{x^2}{5y}$

Solutions:

a. $\log 6x^3 = \log 6 + \log x^3 = \log 6 + 3 \log x$

b. $\ln \dfrac{x^2}{5y} = \ln x^2 - \ln(5y) = 2 \ln x - [\ln 5 + \ln y] = 2 \ln x - \ln 5 - \ln y$

Example 12
Graph each function, state the domain and range, and indicate any asymptotes using dotted lines.

a. $y = \log(x + 2)$

b. $y = 2^x - 2$

Solutions:

a. Use your graphing calculator. Be careful to enclose $x + 2$ in parentheses.

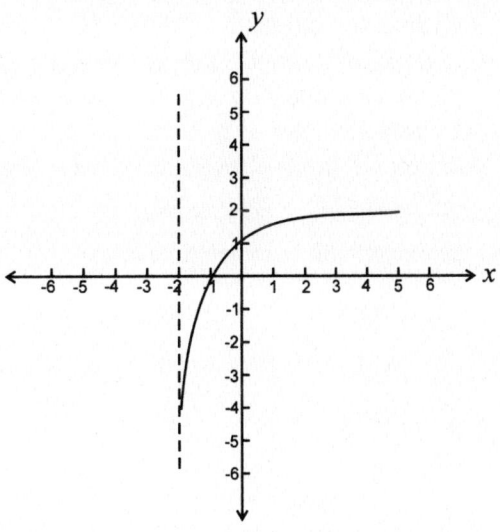

The domain is $(-2, \infty)$, the range is $(-\infty, \infty)$, and there is a vertical asymptote at $x = -2$. Notice the graph of $y = \log(x + 2)$ is the same as the graph of $y = \log x$ shifted 2 units to the left.

b.

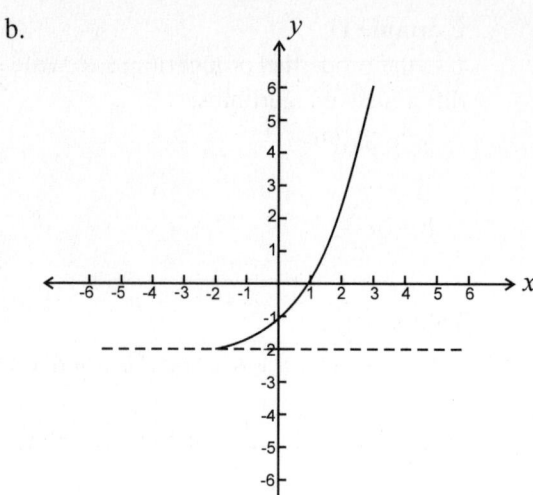

The domain is $(-\infty,\infty)$, the range is $(-2, \infty)$, and there is a horizontal asymptote at $y = -2$. Notice the graph of $y = 2^x - 2$ is the same as the graph of $y = 2^x$ shifted down 2 units.

LIMITS

It is a straightforward process to evaluate a function $y = f(x)$ at a particular x value. If the function is defined at that value, substitute the value for x into $f(x)$ and compute the answer. If the function is not defined at that particular value, f of that value does not exist. However, a limit can exist at a particular value even if the function is not defined at that value. The concept of limits allows one to make note of what y value (if any) is approached as x approaches a particular value.

Some guidelines to help in evaluating $\lim_{x \to c} f(x)$ follow.

1. Given a table of values, if y seems to get closer to the same number as x gets closer to c from the left and right, then $\lim_{x \to c} f(x)$ equals that number. If not, the limit does not exist.

2. Given an expression for $f(x)$, substitute c for x in $f(x)$. If the resulting answer exists, that answer is the limit. If the substitution results in a division by 0, use algebraic techniques to produce an equivalent function which can then be evaluated by substitution. Note that this technique is only appropriate when $f(x)$ is continuous at $x = c$ or when $f(x)$ has a removable discontinuity at $x = c$.

3. Given a graph of $y = f(x)$ (or graphing a given function on your graphing calculator), examine the graph as x approaches c. If the graph appears to approach the same y value from the left and right of c, the limit probably exists and equals the y value.

FUNCTIONS, GRAPHS, AND LIMITS

Example 1

Use the information in the table to find $\lim_{x \to 3} f(x)$, where the table represents values obtained from evaluating $f(x)$ at the given x values.

a.

x	2.9	2.99	2.999	3.001	3.01	3.1
$f(x)$	.2564103	.2506266	.2500625	.2499375	.2493766	.2439024

b.

x	2.9	2.99	2.999	3.001	3.01	3.1
$f(x)$	-10	-100	-1000	1000	100	10

Solutions:

a. As x approaches 3 from the left (2.9, 2.99, 2.999), y gets close to 0.25. As x approaches 3 from the right (3.1, 3.01, 3.001), y gets close to 0.25. The limit appears to be 0.25 or 1/4.

b. As x approaches 3 from the left, the y values are becoming large negative numbers. As x approaches 3 from the right, the y values are becoming large positive numbers. Since the y values do not appear to approach the same number, this limit does not exist.

Example 2

Given the graphs below, determine $\lim_{x \to -2} f(x)$.

a.
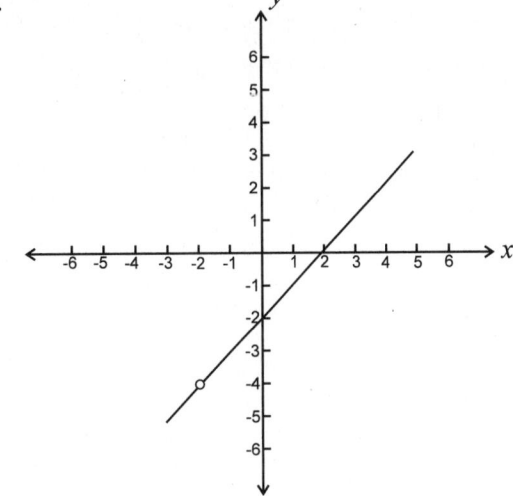

b.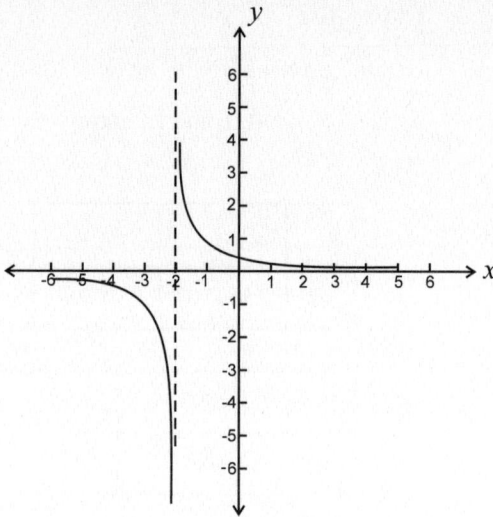

c.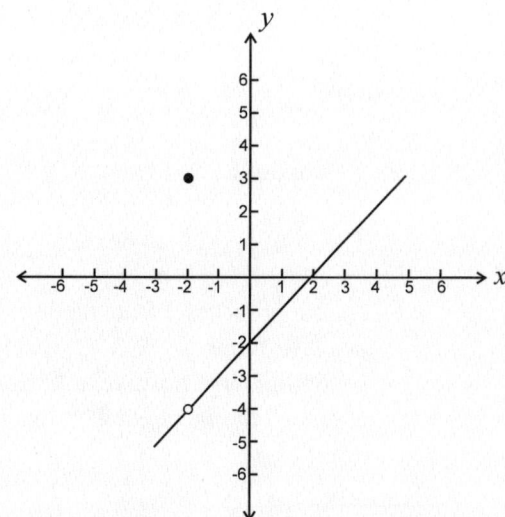

a. $\lim_{x \to -2} f(x) = -4$. Although $f(-2)$ does not exist, (note the open hole below $x = -2$), the limit does exist since y approaches -4 as x approaches -2.

b. $\lim_{x \to -2} f(x)$ does not exist. From the graph, the limit as x approaches -2 from the left decreases without bound. The limit as x approaches -2 from the right increases without bound.

c. $\lim_{x \to -2} f(x) = -1$. Even though , $f(-2) = 3$, the y values are approaching -4 as x approaches -2.

FUNCTIONS, GRAPHS, AND LIMITS

Example 3
Find each limit.

a. $\lim\limits_{x \to 0}(2x^2+1)$

b. $\lim\limits_{x \to \frac{\pi}{2}} \sin 4x$

c. $\lim\limits_{x \to 3}\left(\dfrac{x-3}{x^2-9}\right)$

d. $\lim\limits_{x \to 0} \dfrac{\sqrt{x+9}-3}{x}$

e. $\lim\limits_{\theta \to 0} \dfrac{1-\cos\theta}{4\sin^2\theta}$

Solutions:

a. $\lim\limits_{x \to 0}(2x^2+1) = 2(0)^2 + 1 = 1$

b. $\lim\limits_{x \to \frac{\pi}{2}} \sin 4x = \sin 4\left(\dfrac{\pi}{2}\right) = \sin 2\pi = 0$

c. $\lim\limits_{x \to 3}\left(\dfrac{x-3}{x^2-9}\right) = \dfrac{0}{0}$. Since direct substitution yields division by 0, use factoring first:

$$\lim\limits_{x \to 3}\left(\dfrac{x-3}{x^2-9}\right) = \lim\limits_{x \to 3}\dfrac{x-3}{(x-3)(x+3)} = \lim\limits_{x \to 3}\dfrac{1}{x+3} = \dfrac{1}{3+3} = \dfrac{1}{6}$$

d. $\lim\limits_{x \to 0} \dfrac{\sqrt{x+9}-3}{x} = \dfrac{0}{0}$. Since direct substitution yields division by 0, multiply the numerator and denominator by the conjugate of the numerator:

$$\lim\limits_{x \to 0} \dfrac{\sqrt{x+9}-3}{x} \cdot \dfrac{\sqrt{x+9}+3}{\sqrt{x+9}+3} = \lim\limits_{x \to 0} \dfrac{x+9-9}{x(\sqrt{x+9}+3)} = \lim\limits_{x \to 0} \dfrac{x}{x(\sqrt{x+9}+3)} =$$

$$\lim\limits_{x \to 0} \dfrac{1}{\sqrt{x+9}+3} = \lim\limits_{x \to 0} \dfrac{1}{\sqrt{0+9}+3} = \dfrac{1}{6}$$

e. $\lim\limits_{\theta \to 0} \dfrac{1-\cos\theta}{4\sin^2\theta} = \dfrac{0}{0}$. Use a trigonometric identity ($\sin^2\theta = 1 - \cos^2\theta$) to write this as $\lim\limits_{\theta \to 0} \dfrac{1-\cos\theta}{4(1-\cos^2\theta)}$. Then factor, cancel, and substitute:

$$\lim\limits_{\theta \to 0} \dfrac{1-\cos\theta}{4(1-\cos\theta)(1+\cos\theta)} = \lim\limits_{\theta \to 0} \dfrac{1}{4(1+\cos\theta)} = \dfrac{1}{4(1+1)} = \dfrac{1}{8}$$

Trigonometric Limits

The following trigonometric limits, combined with the use of trigonometric identities, can be used to evaluate limits of trigonometric functions that yield division by 0.

1. $\lim\limits_{x \to 0} \dfrac{\sin x}{x} = 1$

2. $\lim\limits_{x \to 0} \dfrac{1 - \cos x}{x} = 0$

Example 4
Find each limit.

a. $\lim\limits_{x \to \frac{\pi}{2}} \dfrac{\cot x}{\cos x}$

b. $\lim\limits_{x \to 0} \dfrac{\sin 4x}{x}$

c. $\lim\limits_{x \to 0} \dfrac{1 - \cos^2 x}{x}$

Solutions:

a. Substitution yields $\dfrac{0}{0}$. Use the trigonometric identities:

$$\lim_{x \to \frac{\pi}{2}} \frac{\cot x}{\cos x} = \lim_{x \to \frac{\pi}{2}} \frac{\frac{\cos x}{\sin x}}{\frac{\cos x}{1}} =$$

$$\lim_{x \to \frac{\pi}{2}} \frac{\cos x}{\sin x} \cdot \frac{1}{\cos x} = \lim_{x \to \frac{\pi}{2}} \frac{1}{\sin x} = \frac{1}{\sin \frac{\pi}{2}} = 1$$

b. Substitution yields $\dfrac{0}{0}$. Multiply by a form of 1:

$$\lim_{x \to 0} \left(\frac{4}{4} \cdot \frac{\sin 4x}{x} \right) = 4 \lim_{x \to 0} \left(\frac{\sin 4x}{4x} \right) = 4(1) = 4.$$

c. Substituting 0 for x will result in division by 0. Use factoring to write

$$\lim_{x \to 0} \frac{1 - \cos^2 x}{x} =$$

FUNCTIONS, GRAPHS, AND LIMITS

$$\lim_{x \to 0} \frac{(1-\cos x)(1+\cos x)}{x} = \lim_{x \to 0} \frac{(1-\cos x)}{x} \cdot (1+\cos x) = 0 \cdot 2 = 0.$$

ONE-SIDED LIMITS

The previous limits involved evaluating functions as x approached a value from both the left and right side. The same rules stated previously apply to one-sided limits: try substitution, if that fails (by yielding division by 0), use algebraic techniques and/or a graph to help evaluate the limit.

Example 5
Evaluate each limit.

a. $\lim\limits_{x \to 0^-}(x^2 - 4)$

b. $\lim\limits_{x \to 5^+} \dfrac{x^2 - 25}{x - 5}$

c. $\lim\limits_{x \to 2^+} \dfrac{|x - 2|}{x - 2}$

d. $\lim\limits_{x \to 3^-} \sqrt{x - 3}$

Solutions:

a. $\lim\limits_{x \to 0^-}(x^2 - 4) = 0^2 - 4 = -4$

b. $\lim\limits_{x \to 5^+} \dfrac{x^2 - 25}{x - 5} = \lim\limits_{x \to 5^+} \dfrac{(x-5)(x+5)}{x - 5} = \lim\limits_{x \to 5^+}(x+5) = 5 + 5 = 10$

c. $\lim\limits_{x \to 2^+} \dfrac{|x - 2|}{x - 2} = \lim\limits_{x \to 2^+} \dfrac{x-2}{x-2} = 1$ since $x - 2 > 0$ as x approaches 2 from the right.

An alternate approach is to examine the graph of $y = \dfrac{|x-2|}{x-2}$:

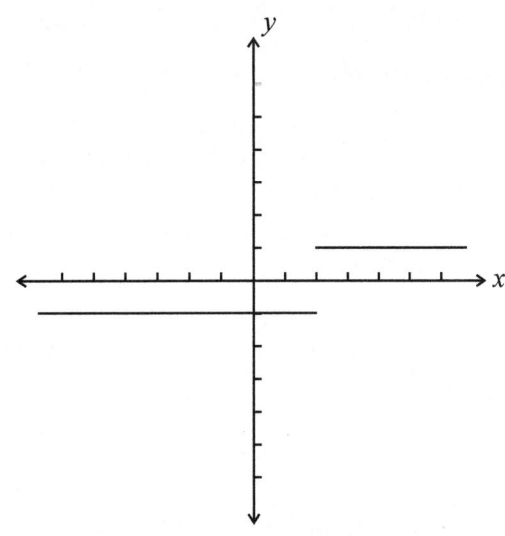

Note that from the right, $y = 1$, and thus $\lim\limits_{x \to 2^+} \dfrac{|x-2|}{x-2} = 1$. Note also that $\lim\limits_{x \to 2^-} \dfrac{|x-2|}{x-2} = -1$, and that $\lim\limits_{x \to 2} \dfrac{|x-2|}{x-2}$ does not exist. This demonstrates the concept that a limit as x approaches c exists if and only if the limit from the left and right of c exist and are equal.

d. $\lim\limits_{x \to 3^-} \sqrt{x-3}$ does not exist. Substitution would give an answer of 0, which is not correct. Examine the graph:

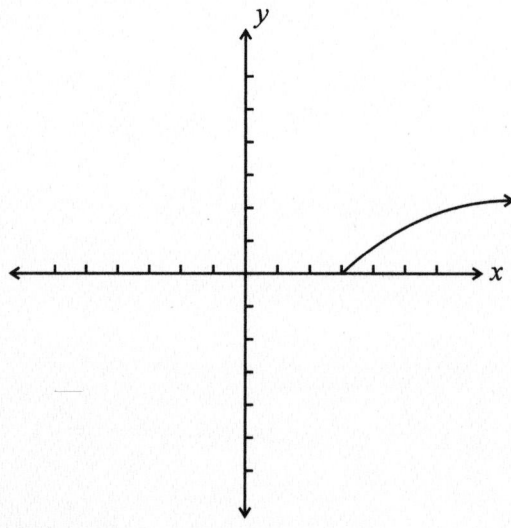

and notice that the graph does not exist as x approaches 3 from the left. Hence, the limit does not exist.

ASYMPTOTIC AND UNBOUNDED BEHAVIOR

Asymptotes occur in several types of functions including rational functions, exponential, and logarithmic functions. If the graph increases (or decreases) without bound, although the limit does not exist, we use positive or negative infinity to represent what the graph is doing. Consider the following possibilities.

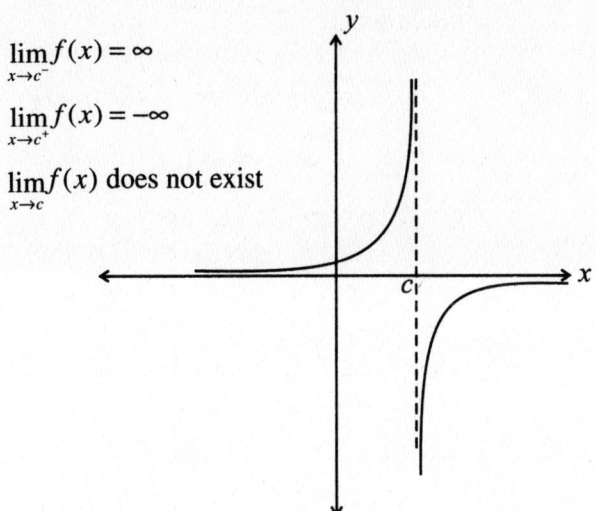

$\lim\limits_{x \to c^-} f(x) = \infty$

$\lim\limits_{x \to c^+} f(x) = -\infty$

$\lim\limits_{x \to c} f(x)$ does not exist

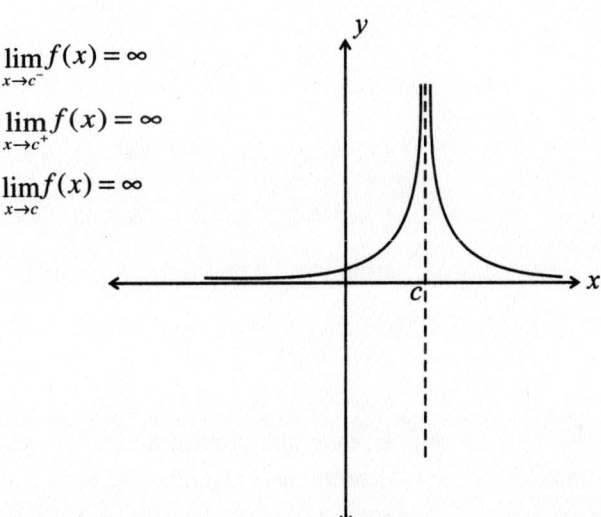

$\lim\limits_{x \to c^-} f(x) = \infty$

$\lim\limits_{x \to c^+} f(x) = \infty$

$\lim\limits_{x \to c} f(x) = \infty$

Example 6
Find each limit.

a. $\lim\limits_{x \to 2^+} \dfrac{1}{x-2}$

b. $\lim\limits_{x \to 0^-} \dfrac{1}{x}$

c. $\lim\limits_{x \to \frac{\pi}{4}} \tan 2x$

d. $\lim\limits_{x \to 0^+} \dfrac{1}{\sin x}$

PART I

Solutions:

a. Using a graphing calculator to graph $y = \dfrac{1}{x-2}$:

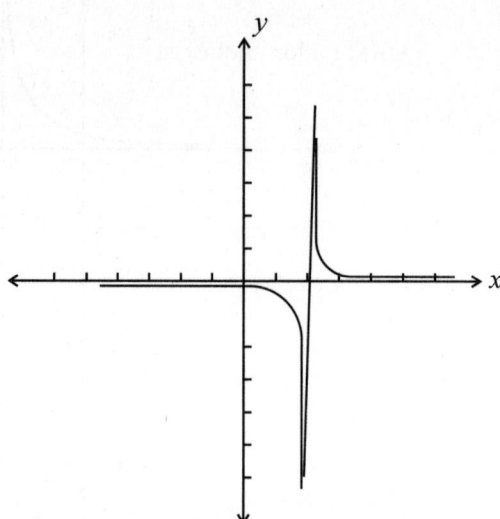

In connect dot mode, the calculator appears to show the asymptote, when actually it is connecting the computed values as x approaches 2 from the left and right. As x approaches 2 from the right, the graph increases without bound and $\lim\limits_{x \to 2^+} \dfrac{1}{x-2} = +\infty$.

b. Use a graph or reason as follows. As x approaches 0 from the left, the denominator is a small negative number. The numerator is positive, and since a positive divided by a negative is negative, the limit approaches $-\infty$.

c. Using a graph of $y = \tan 2x$:

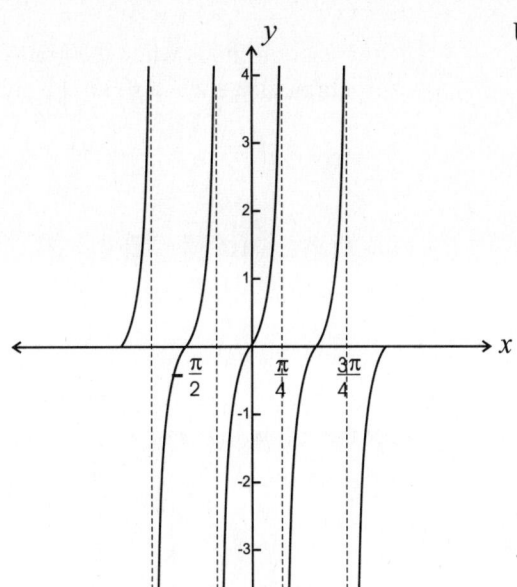

d. Using a graph of $y = \dfrac{1}{\sin x}$ (done in connect mode, standard viewing window): $\lim\limits_{x \to \left(\frac{\pi}{4}\right)^-} \tan 2x = +\infty$, $\lim\limits_{x \to \left(\frac{\pi}{4}\right)^+} \tan 2x = -\infty$, so $\lim\limits_{x \to \frac{\pi}{4}} \tan 2x$ does not exist.

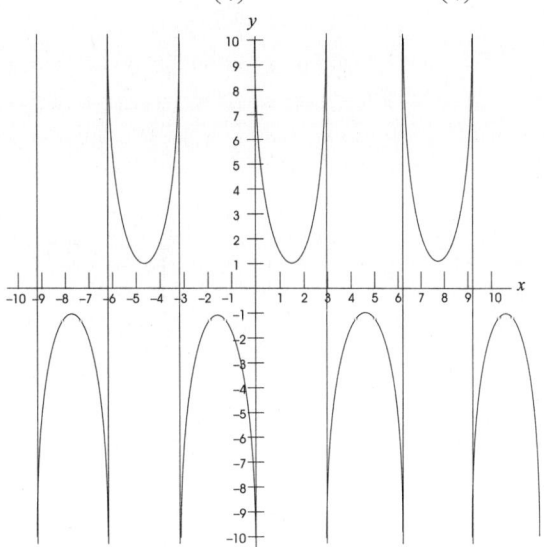

$\lim\limits_{x \to 0^+} \dfrac{1}{\sin x} = +\infty$

CONTINUITY

A function is continuous when the graph can be drawn without breaks or holes. Algebraically, for $y = f(x)$ to be continuous at $x = a$ means:

1. $f(a)$ exists.

2. $\lim_{x \to a} f(x)$ exists

3. $\lim_{x \to a} f(x) = f(a)$

Consider the following graphs.

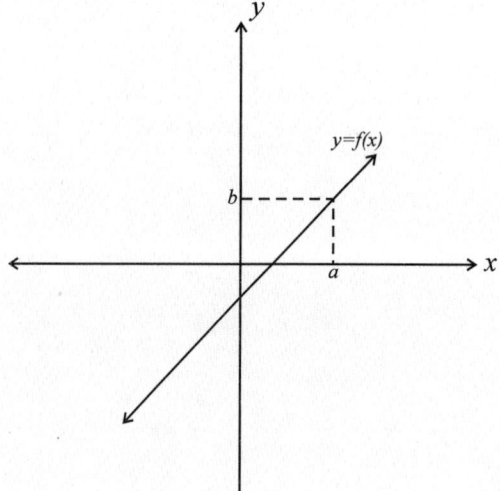

$f(a) = b$

$\lim_{x \to a} f(x) = b$

$\lim_{x \to a} f(x) = f(a) = b$

Therefore, $y = f(x)$ is continuous at $x = a$

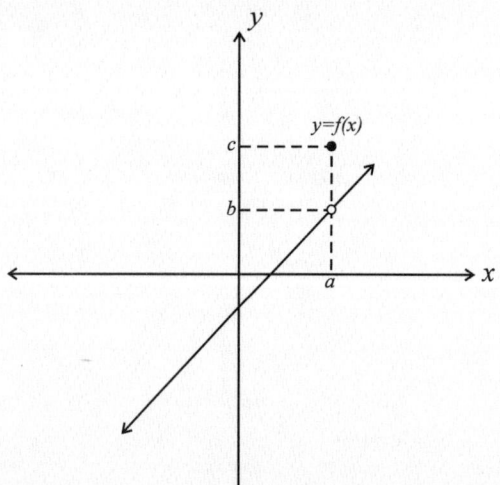

$f(a) = c$

$\lim_{x \to a} f(x) = b$

$\lim_{x \to a} f(x) \neq f(a)$

Therefore, $y = f(x)$ is not continuous at $x = a$

FUNCTIONS, GRAPHS, AND LIMITS

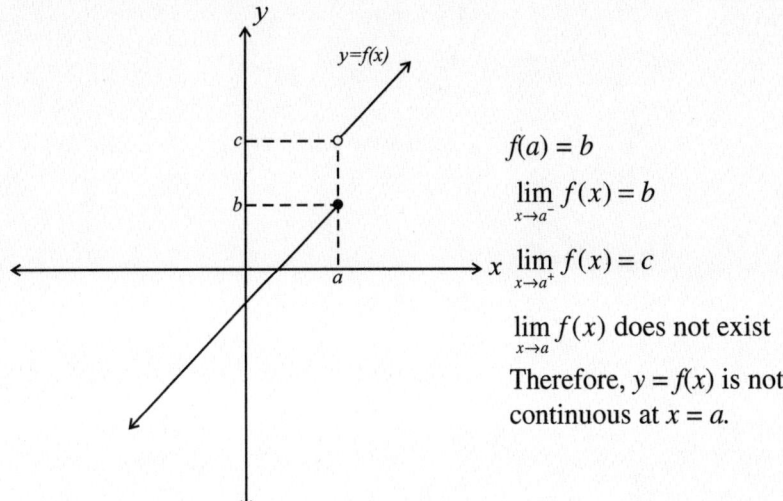

$f(a) = b$

$\lim_{x \to a^-} f(x) = b$

$\lim_{x \to a^+} f(x) = c$

$\lim_{x \to a} f(x)$ does not exist

Therefore, $y = f(x)$ is not continuous at $x = a$.

Example 7

Determine whether the function $f(x) = \dfrac{x^2 - 4}{x - 2}$ is continuous at $x = 2$.

Solution:

Since $x = 2$ is not in the domain of the function, $f(2)$ does not exist and the function cannot be continuous at $x = 2$. Use care with your graphing calculator as it appears in a standard viewing window that the function is continuous:

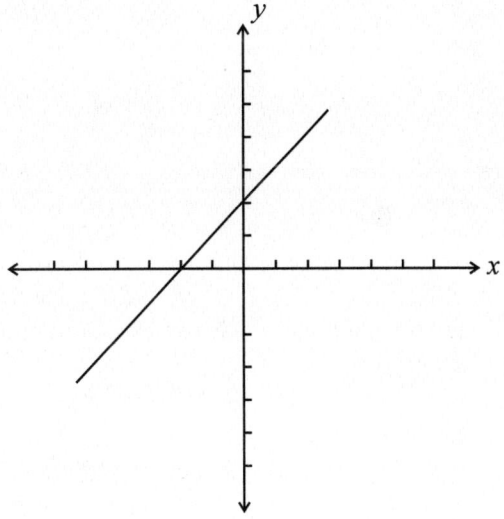

However, since $x = 2$ is not in the domain, you should realize that there is a hole at $x = 2$.

Example 8

Determine whether the function $f(x) = \begin{cases} \dfrac{x^2-4}{x-2}, & x \neq 2 \\ 4, & x = 2 \end{cases}$ is continuous at $x = 2$.

Solution:

$x = 2$ is now in the domain of f, and $f(2) = 4$.

$$\lim_{x \to 2} \frac{x^2-4}{x-2} = \lim_{x \to 2} \frac{(x-2)(x+2)}{x-2} = \lim_{x \to 2}(x+2) = 4.$$

Thus f is continuous at $x = 2$.

Example 9

At what points, if any, is the function $f(x) = \begin{cases} 2x, & x < 1 \\ 3x+4, & x \geq 1 \end{cases}$ discontinuous?

Solution:

The two pieces are continuous on their domains (each is a line). Thus, you need only check $x = 1$.

$\lim_{x \to 1^+} f(x) = 2(1) = 2$; $\lim_{x \to 1^-} f(x) = 3(1) + 4 + 7$. Since the left and right hand limits are not equal, the limit does not exist and f is discontinuous at $x = 1$.

INTERMEDIATE VALUE THEOREM

The Intermediate Value Theorem states that if f is continuous on the closed interval $[a, b]$ and C is any number such that $f(a) \leq C \leq f(b)$, then there exists at least one number c where $a \leq c \leq b$ such that $f(c) = C$. Consider the following example.

Example 10

If $f(x) = x^2 + 2x + 1$, then by the Intermediate Value Theorem there exists at least one c in the interval $-2 \leq x \leq 2$ that satisfies $f(c) = 1$. Find c.

Solution:

Solving for c gives $c^2 + 2c = 0$, $c(c + 2) = 0$, so $c = 0$ or $c = -2$. In this instance, we found two values for c.

FUNCTIONS, GRAPHS, AND LIMITS

Example 11
Give a written explanation of why the function $g(x) = x^2 + 4x + 3$ has a zero in the interval $[-4, 2]$.

Solution:

$g(-4) = (-4)^2 + 4(-4) + 3 = 3$ and $g(-2) = (-2)^2 + 4(-2) + 3 = -1$. By the Intermediate Value Theorem there exists at least one c between -4 and -2 where $g(c) = 0$, since 0 is between -1 and 3. That value (or these values) is the zero of the function.

There is a corollary to the Intermediate Value Theorem known as the Extreme Value Theorem. This theorem is used to determine a maximum and/or minimum value on a *closed interval*. The Extreme Value Theorem states:

> If f is continuous on a closed interval $[a,b]$, then f has both a minimum and a maximum in the interval.

These minimum and maximum values are referred to as *extrema*. Extrema that occur at the endpoints of an interval are called *end-point extrema*. Extrema that are found interior to the interval are known as *relative extrema*.

Something to bear in mind when dealing with extrema is that a function may not have either a minimum or maximum value. There is even the possibility that neither exists.

In the first graph, (a), we see a closed interval that has both a minimum and a maximum. More, the minimum is an example of relative extrema and the maximum is an end-point extrema.

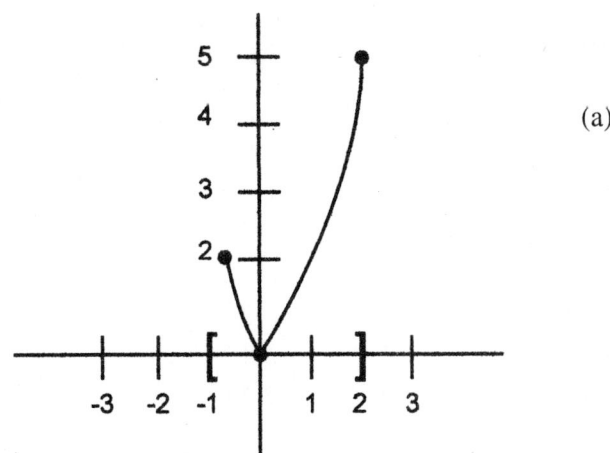

(a)

PART I

Graph a: end-point extrema at (−1,2) and (2,5). Relative extrema at (0,0) on the closed interval whose domain is [−1,2]. The minimum exists at (0,0) and the maximum for the interval is at (2,5).

In graph b, though f is continuous, it is an open interval (−1,2) where there is a minimum at $x = 0$ but no maximum.

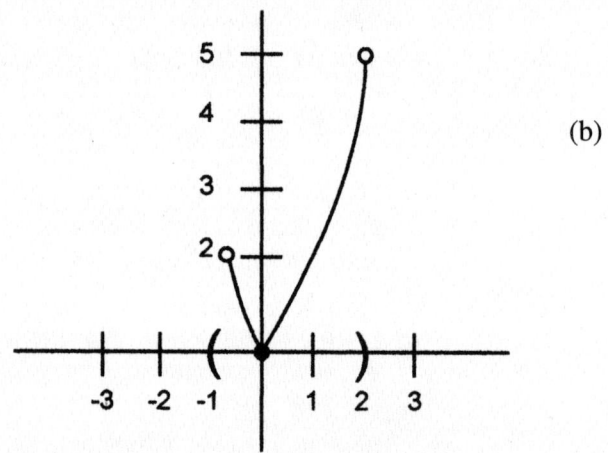

(b)

If we were to graph $f(x) = (3x^2 - 2)/5$ on the open interval whose domain is (−3,2), we get:

$$f(-3) = \frac{\left(3(-3)^2 - 2\right)}{5}$$
$$= (3(9) - 2)/5$$
$$= (27 - 2)/5$$
$$= 25/5$$
$$= 5$$

$$f(2) = \frac{\left(3(2)^2 - 2\right)}{5}$$
$$= (3(4) - 2)/5$$
$$= (12 - 2)/5$$
$$= 10/5$$
$$= 2$$

Graph c shows the resulting plot of these values on the open interval. Note the relative extrema at (0,0). This is the minimum for f. Note also that there is no maximum.

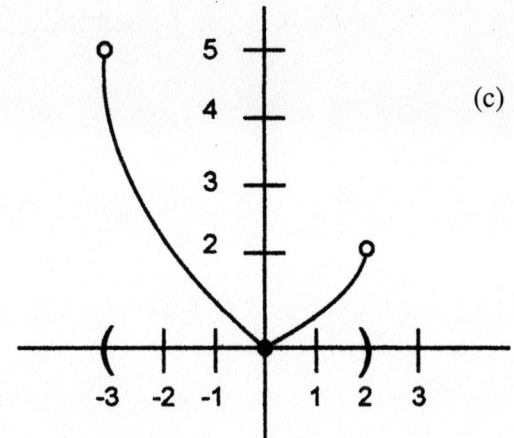
(c)

Part II
DERIVATIVES

Definition of Derivatives

The slope of a line can be found using $m = \dfrac{y_2 - y_1}{x_2 - x_1}$ if two points (x_1, y_1) and (x_2, y_2) are known. Using function notation, the points can be written as $(x, f(x))$ and $(x+h, f(x+h))$ and slope can be written as

$$m = \dfrac{f(x+h) - f(x)}{x+h-x} = \dfrac{f(x-h) - f(x)}{h}$$

This formula represents the average rate of change of the function over the interval $[x, x+h]$. When we want to find the slope of the tangent line or instantaneous rate of change, we let h approach 0, and define the derivative of f at x as $f'(x) = \lim\limits_{h \to 0} \dfrac{f(x=h) - f(x)}{h}$ if this limit exists.

Example 1
Use the graph to estimate the slope of the line from the given graph.

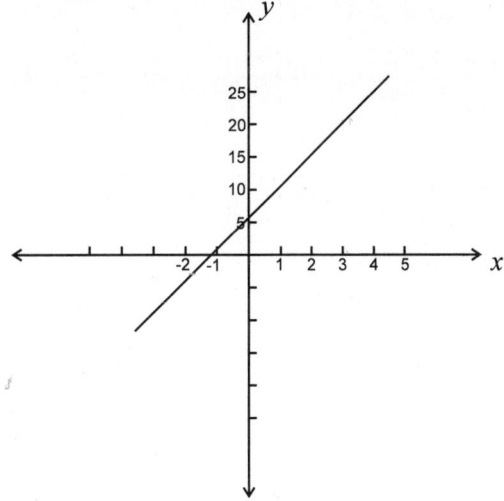

Solution:
Use two points on the line, such as (0, 5) and (3, 20) to find the slope.

$$m = \dfrac{y_2 - y_1}{x_2 - x_1} = \dfrac{20-5}{3-0} = \dfrac{15}{3} = 5$$

PART II

Example 2

Use the graph to estimate the slope of the curve at the point (x, y).

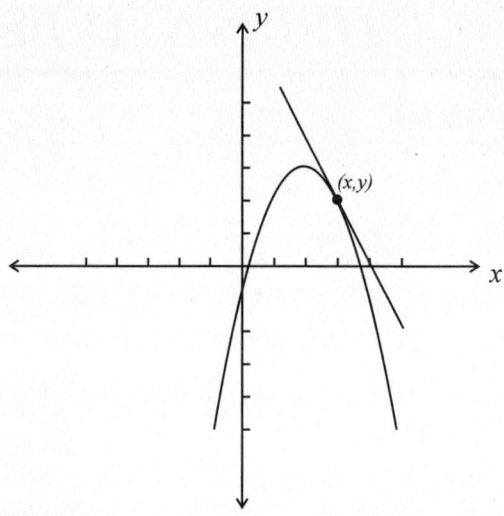

Solution:

Assuming the scales on the axes are equal and starting at the point (x, y), the line falls 2 units and runs to the right 1 unit, giving a slope of the tangent of –2.

Example 3

The table below gives the distance traveled in inches after t seconds of an object traveling along a line.

a. Find the average velocity of the object from $t = 0$ to $t = 3$ seconds.

b. Find the average velocity of the object from $t = 2$ to $t = 3$ seconds.

t	0	1	2	3	4
$s(t)$	1	2	5	10	17

Solution:

a. Average velocity = $\dfrac{\text{change in distance}}{\text{change in time}} = \dfrac{\Delta s}{\Delta t} = \dfrac{10-1}{3-0} = \dfrac{9}{3} = 3$ in/sec

b. $\dfrac{\Delta s}{\Delta t} = \dfrac{10-5}{3-2} = \dfrac{5}{1} = 5$ in/sec

DERIVATIVES

Example 4

An object travels along a line so that its distance traveled in inches after t seconds is $s(t) = \sqrt{2t+1}$.

   a. Find the average velocity over the interval [0, 5].

   b. Find the instantaneous velocity after 5 seconds.

Solutions:

   a. $s(5) = \sqrt{2(5)+1} = \sqrt{11}$.

      $s(0) = \sqrt{2(0)+1} = \sqrt{1} = 1$.

      Average velocity =

$$\frac{\text{change in distance}}{\text{change in time}} = \frac{\sqrt{11}-1}{5-0} = \frac{\sqrt{11}-1}{5} \approx 0.46 \text{ inches/second.}$$

   b. To find instantaneous velocity at $t = 5$, we find

$$\lim_{h \to 0} \frac{s(5+h)-s(5)}{h} =$$

$$\lim_{h \to 0} \frac{\sqrt{2(5+h)+1}-\sqrt{2(5)+1}}{h} = \lim_{h \to 0} \frac{\sqrt{11+2h}-\sqrt{11}}{h} \cdot \frac{\sqrt{11+2h}+\sqrt{11}}{\sqrt{11+2h}+\sqrt{11}} =$$

$$\lim_{h \to 0} \frac{11+2h-11}{h(\sqrt{11+2h}+\sqrt{11})} = \lim_{h \to 0} \frac{2h}{h(\sqrt{11+2h}+\sqrt{11})} = \lim_{h \to 0} \frac{2}{\sqrt{11+2h}+\sqrt{11}} =$$

$$\frac{2}{\sqrt{11}+\sqrt{11}} = \frac{2}{2\sqrt{11}} = \frac{1}{\sqrt{11}} \approx 0.30 \text{ in/sec}$$

ALTERNATE FORM OF THE DEFINITION OF DERIVATIVE

An alternate form of the limit definition of a derivative of f at a is given by:

$$f'(a) = \lim_{x \to a} \frac{f(x)-f(a)}{x-a}, \text{ provided this limit exists.}$$

Example 5

Use $\lim_{h \to 0} \frac{f(x+h)-f(x)}{h}$ and $\lim_{x \to a} \frac{f(x)-f(a)}{x-a}$ to find $f'(a)$ if $f(x) = 3x^2$.

PART II

Solution:

$$\lim_{h \to 0} \frac{f(x+h)-f(x)}{h} = \lim_{h \to 0} \frac{3(a+h)^2 - 3a^2}{h} = \lim_{h \to 0} \frac{3a^2 + 6ah + 3h^2 - 3a^2}{h} =$$

$$\lim_{h \to 0} \frac{h(6a+3h)}{h} = \lim_{h \to 0} 6a + 3h = 6a$$

$$\lim_{x \to a} \frac{f(x)-f(a)}{x-a} = \lim_{x \to a} \frac{3x^2 - 3a^2}{x-a} = \lim_{x \to a} \frac{3(x^2-a^2)}{x-a} = \lim_{x \to a} \frac{3(x+a)(x-a)}{x-a} =$$

$$\lim_{x \to a} 3(x+a) = 6a$$

DIFFERENTIABILITY AND CONTINUITY

Continuity was discussed previously (see Chapter 1). The following statements summarize the relationship between differentiability and continuity.

1. If f is not continuous at $x = c$, f is not differentiable at $x = c$.
2. If f is differentiable at $x = c$, f is continuous at $x = c$.
   (Differentiability implies continuity.)
3. If f is continuous at $x = c$, f may or may not be differentiable at $x = c$.
   (Continuity does not imply differentiability.)

Example 6
Which functions are continuous at $x = 0$? Which functions are differentiable at $x = 0$?

a. $f(x) = |x|$ b. $f(x) = \sqrt[3]{x}$ c. $f(x) = [x]$

Solutions:

a. $f(x) = |x|$ is continuous at $x = 0$. It is *not* differentiable at $x = 0$. Consider the one-sided limits using the alternate form of the limit definition of derivative:

$$\lim_{x \to 0^-} \frac{f(x)-f(0)}{x-0} = \lim_{x \to 0^-} \frac{|x|-|0|}{x-0} = \lim_{x \to 0^-} \frac{|x|}{x} = -1$$

$$\lim_{x \to 0^+} \frac{|x|-|0|}{x-0} = \lim_{x \to 0^+} \frac{|x|}{x} = +1$$

DERIVATIVES

Since the limit from the left does not equal the limit from the right, the limit as x approaches 0 does not exist, and the derivative does not exist.

b. A graphing calculator will help you verify that $f(x) = \sqrt[3]{x}$ is continuous at $x = 0$. However, checking the limit, we find:

$$\lim_{x \to 0} \frac{f(x) - f(0)}{x - 0} = \lim_{x \to 0} \frac{\sqrt[3]{x}}{x} = \lim_{x \to 0} \frac{1}{x^{2/3}} = \infty$$

so the function is not differentiable at $x = 0$.

c. The greatest integer function is not continuous at $x = 0$, and therefore is not differentiable at $x = 0$.

BASIC RULES OF DIFFERENTIATION

A summary of the basic rules of differentiation follows. These rules must be memorized.

1. The derivative of a constant equals 0. $\frac{d}{dx}[c] = 0$

2. The derivative of x^n is nx^{n-1}. $\frac{d}{dx}[x^n] = nx^{n-1}$

3. The derivative of a constant times a function is the constant times the derivative of the function.

   $$\frac{d}{dx}[cf(x)] = c\frac{d}{dx}[f(x)]$$

4. The derivative of a sum or difference is the sum or difference of the derivatives.

   $$\frac{d}{dx}[f(x) \pm g(x)] = \frac{d}{dx}[f(x)] \pm \frac{d}{dx}[g(x)]$$

Example 7
Find the derivative of each function.

a. $f(x) = 4x^3 - 7x^2 + 2$ b. $g(x) = \sqrt[3]{x} + \frac{5}{\sqrt{x}}$ c. $s(t) = \frac{1}{(5t)^2}$

Solutions:

a. $f'(x) = 4(3x^2) - 7(2x) + 0 = 12x^2 - 14x$

b. Rewrite $g(x)$ first as

$g(x) = x^{1/3} + 5x^{-1/2}$. Then $g'(x) = \dfrac{1}{3}x^{-2/3} - \dfrac{5}{2}x^{-3/2}$

c. Rewrite $s(t) = \dfrac{1}{25t^2} = \dfrac{1}{25}t^{-2}$. Then $s'(t) = \dfrac{1}{25}(-2t^{-3}) = -\dfrac{2}{25}t^{-3} = -\dfrac{2}{25t^3}$

5. The Product Rule is $\dfrac{d}{dx}[f(x) \cdot g(x)] = f(x) \cdot g'(x) + g(x) \cdot f'(x)$.

6. The Quotient Rule is $\dfrac{d}{dx}\left[\dfrac{f(x)}{g(x)}\right] = \dfrac{g(x) \cdot f'(x) - f(x) \cdot g'(x)}{[g(x)]^2}$

Example 8
Find the derivative of each function.

a. $h(x) = (x^2 + 1)(2x + 3)$ b. $g(x) = \dfrac{2x+1}{4x-3}$

Solutions:

a. $h'(x) = (x^2 + 1)(2) + (2x + 3)(2x) = 2x^2 + 2 + 4x^2 + 6x = 6x^2 + 6x + 2$

This can be checked by first multiplying out $h(x) = 2x^3 + 3x^2 + 2x + 3$

and using the previously stated rules: $h'(x) = 6x^2 + 6x + 2$.

b. $g'(x) = \dfrac{(4x-3)(2) - (2x+1)(4)}{(4x-3)^2} = \dfrac{8x - 6 - 8x - 4}{(4x-3)^2} = \dfrac{10}{(4x-3)^2}$

7. The derivatives of the six trigonometric functions are:

$\dfrac{d}{dx}[\sin x] = \cos x$ $\dfrac{d}{dx}[\csc x] = -\csc x \cot x$

$\dfrac{d}{dx}[\cos x] = -\sin x$ $\dfrac{d}{dx}[\sec x] = \sec x \tan x$

$\dfrac{d}{dx}[\tan x] = \sec^2 x$ $\dfrac{d}{dx}[\cot x] = -\csc^2 x$

DERIVATIVES

8. The derivatives of the six inverse trigonometric functions are:

$$\frac{d}{dx}[\arcsin x] = \frac{1}{\sqrt{1-x^2}} \qquad \frac{d}{dx}[\operatorname{arccsc} x] = \frac{-1}{|x|\sqrt{x^2-1}}$$

$$\frac{d}{dx}[\arccos x] = \frac{-1}{\sqrt{1-x^2}} \qquad \frac{d}{dx}[\operatorname{arcsec} x] = \frac{1}{|x|\sqrt{x^2-1}}$$

$$\frac{d}{dx}[\arctan x] = \frac{1}{1+x^2} \qquad \frac{d}{dx}[\operatorname{arccot} x] = \frac{-1}{1+x^2}$$

9. The derivatives of the logarithmic and exponential functions are:

$$\frac{d}{dx}[\ln x] = \frac{1}{x} \qquad \frac{d}{dx}[e^x] = e^x$$

$$\frac{d}{dx}[\log_a x] = \frac{1}{x \ln a} \qquad \frac{d}{dx}[a^x] = a^x \ln a$$

Example 9
Find the derivative of each function.

a. $y = \sin x + \tan x$ b. $y = \dfrac{\sin x}{\cos x}$

c. $y = \arccos x - \cos x$ d. $y = e^x \ln x$

Solutions:

a. $y' = \cos x + \sec^2 x$ by adding the derivative of each term.

b. Using the quotient rule:

$$y' = \frac{\cos x(\cos x) - \sin x(-\sin x)}{(\cos x)^2} = \frac{\cos^2 x + \sin^2 x}{\cos^2 x} = \frac{1}{\cos^2 x} = \sec^2 x.$$

Or, by using a trigonometric identity first, and writing

$$y = \frac{\sin x}{\cos x} = \tan x \text{ and a rule: } y' = \sec^2 x.$$

c. $y' = \dfrac{-1}{\sqrt{1-x^2}} - (-\sin x) = \dfrac{-1}{\sqrt{1-x^2}} + \sin x$

d. Using the product rule:

$$y' = e^x\left(\frac{1}{x}\right) + (\ln x)e^x = e^x\left(\frac{1}{x} + \ln x\right) = e^x\left(\frac{1+x\ln x}{x}\right)$$

Note that the first step was a calculus step, but the last two steps were algebraic simplification.

CHAIN RULE

The formulas for derivatives were stated as functions of x. If a function h can be viewed as a composition of functions f and g, the derivative is found using the Chain Rule:

10. If $h(x) = f(g(x))$, then $h'(x) = f'(g(x)) \cdot g'(x)$.

Example 10
Find the derivative of each function.

a. $y = (2x+3)^3$ b. $y = \sqrt{4-x^2}$ c. $y = 4x(2x+3)^3$

Solutions:

a. Consider this function as the composition of $f(x) = x^3$ and $g(x) = 2x+3$. Then $y = f(g(x))$ and $y' = f'(g(x)) \cdot g'(x) = 3(g(x))^2 \cdot 2 = 3(2x+3)^2 \cdot 2 = 6(2x+3)^2$

b. $y = \sqrt{4-x^2} = (4-x^2)^{\frac{1}{2}}$ so
$$y' = \frac{1}{2}(4-x^2)^{-\frac{1}{2}}(-2x) = -x(4-x^2)^{-\frac{1}{2}} = \frac{-x}{\sqrt{4-x^2}}$$

c. The derivative of this function requires use of both the product rule and the chain rule: $y' = 4x[6(2x+3)^2] + (2x+3)^3(4)$ where the expression in the brackets was the derivative obtained in part (a) of this example. Simplify by factoring:
$$y' = 4(2x+3)^2[x \cdot 6 + (2x+3)] = 4(2x+3)^2(8x+3)$$

Example 11
Find the derivative of each function.

a. $y = \sin 6x$ b. $y = 3\tan \pi x$ c. $y = \cos x^2$

d. $y = (\cos x)^2$

Solutions:

a. Consider this function as the composition of $f(x) = \sin x$ and $g(x) = 6x$. Then $y = f(g(x))$ and
$y' = f'(g(x)) \cdot g'(x) = \cos(g(x)) \cdot 6 = \cos(6x) \cdot 6 = 6\cos 6x$

b. $y = 3(\tan(\pi x))$ so $y' = 3(\sec^2(\pi x)) \cdot \pi = 3\pi \sec^2 \pi x$

c. Here, $g(x) = x^2$ and $g'(x) = 2x$ so $y' = -\sin x^2(2x) = -2x \sin x^2$

d. Here $f(x) = x^2$ and $g(x) = \cos x \cdot y' = 2(\cos x) \cdot (-\sin x) =$ $-2 \sin x \cos x$ or $-2 \sin 2x$ using a double angle identity.

Example 12
Find each derivative.

a. $f(x) = \arcsin(4x)$ b. $y = \arctan(x^2)$ c. $g(x) = \ln(2x+1)$

d. $y = e^{2x}$

Solutions:

a. $f'(x) = \dfrac{1}{\sqrt{1-(4x)^2}} \cdot 4 = \dfrac{4}{\sqrt{1-16x^2}}$

b. $y' = \dfrac{1}{1+(x^2)^2} \cdot 2x = \dfrac{2x}{1+x^4}$

c. $g'(x) = \dfrac{1}{2x+1} \cdot 2 = \dfrac{2}{2x+1}$

d. $y' = e^{2x}(2) = 2e^{2x}$

HIGHER ORDER DERIVATIVES

Higher order derivatives are found by taking derivatives of derivatives. You should be familiar with the notation associated with derivatives:

First derivative	$f'(x)$	y'	$D_x(y)$	$\dfrac{d}{dx}[f(x)]$
Second derivative	$f''(x)$	y''	$D^2_x(y)$	$\dfrac{d^2}{dx^2}[f(x)]$
Third derivative	$f'''(x)$	y'''	$D^3_x(y)$	$\dfrac{d^3}{dx^3}[f(x)]$
$\vdots$	$\vdots$	$\vdots$	$\vdots$	$\vdots$
nth derivative	$f^{(n)}(x)$	$y^{(n)}$	$D^{(n)}_x(y)$	$\dfrac{d^n}{dx^n}[f(x)]$

Example 13

Find $\dfrac{dy}{dx}$ and $\dfrac{d^2y}{dx^2}$ for each function.

a. $y = 4x^3 - 6x^2 + 2x + 1$ b. $y = 4x(x^2 - 1)^3$

c. $y = \sin^2(3x)$

Solutions:

a. $\dfrac{dy}{dx} = 12x^2 - 12x + 2;\ \dfrac{d^2y}{dx^2} = 24x - 12$

b. The first derivative is found using the product rule:

$\dfrac{dy}{dx} = 4x\left[3(x^2-1)^2(2x)\right] + (x^2-1)^3(4)$. Simplify by factoring before attempting to take the second derivative:

$\dfrac{dy}{dx} = 4(x^2-1)^2\left[6x^2 + (x^2-1)\right] = 4(x^2-1)^2(7x^2-1)$.

Now, $\dfrac{d^2y}{dx^2} = 4(x^2-1)^2(14x) + (7x^2-1)8(x^2-1)(2x)$. Simplify by factoring:

$\dfrac{d^2y}{dx^2} = 8x(x^2-1)\left[7(x^2-1) + (7x^2-(2))\right] = 8x(x^2-1)(21x^2-9)$

c. It might help to rewrite $y = (\sin(3x))^2$ and then use the chain rule to find the first derivative:

$\dfrac{dy}{dx} = 2(\sin(3x))(\cos(3x)(3)) = 6\sin 3x \cos 3x$. Use the product rule to find the second derivative:

$\dfrac{d^2y}{dx^2} = 6\sin 3x(-3\sin 3x) + \cos 3x(18\cos 3x) = -18\sin^2 3x + 18\cos^2 3x$

or $18\cos(6x)$ using a double angle identity.

Implicit Differentiation

When a function is written implicitly (not solved for y), treat x and y as functions. Some basic steps are:

DERIVATIVES

1. Find the derivative of each term.
2. Collect all terms involving y' on the left side of the equation and all other terms on the right side of the equation.
3. Factor out y' on the left side of the equation.
4. Divide both sides of the equation by the coefficient of y'.

Example 14

Find $\dfrac{dy}{dx}$ if $x^2 + 4xy + y^2 = 8$.

Solution:

Consider the derivative of each term, using the product rule to find the derivative of $4xy$:

$$2x + 4x(y') + y(4) + 2yy' = 0$$
$$4xy' + 2yy' = -2x - 4y$$
$$y'(4x + 2y) = -2x - 4y$$
$$y' = \frac{-2x - 4y}{4x + 2y} = \frac{-2(x + 2y)}{2(2x + y)} = \frac{-(x + 2y)}{2x + y}$$

Example 15

Find $\dfrac{dy}{dx}$ if $x \sin y = 1$.

Solution:

Use the product rule to find $D_x[x \sin y]$:

$$x[\cos y \cdot y'] + \sin y (1) = 0$$
$$x[\cos y \cdot y'] = -\sin y$$
$$y' = \frac{-\sin y}{x \cos y} = -\frac{1}{x} \tan y$$

INVERSE FUNCTIONS

When a function is described by a set of ordered pairs, its inverse can be found easily by interchanging the first and second coordinates of each pair.

Recall from precalculus that if a function f is one-to-one, then its inverse f^{-1} satisfies $f(f^{-1}(x)) = x$ and $f^{-1}(f(x)) = x$.

PART II

Some tips and strategies for problems involving inverse functions follow:

1. Only one-to-one functions have inverses. One-to-one functions pass the horizontal line test, that is, any horizontal line passes through at most one point on the graph of a one-to-one function.

2. Given the equation of a one-to-one function, you can find its inverse by interchanging x and y and solving for y.

3. The graph of $y = f^{-1}$ is a reflection of the graph of $y = f(x)$ in the line $y = x$. Thus, if (a, b) is a point on the graph of $y = f(x)$, then (b, a) is a point on the graph of $y = f^{-1}(x)$.

Example 16
Find f^{-1} if $f(x) = 3x + 1$.

Solution:
Given $y = 3x + 1$ (using y instead of $f(x)$ keeps notation easier), interchange x and y so that $x = 3y + 1$. Then $y = \dfrac{x-1}{3}$ gives the inverse, or

$$f^{-1}(x) = \dfrac{x-1}{3}.$$

DERIVATIVE OF AN INVERSE FUNCTION

In the previous section, it was noted that the graphs of f and f^{-1} are reflections in the line $y = x$. In addition if f and g are inverses of each other and (a, b) is a point on the graph of f (so that (b, a) must be a point on the graph of g), the slope of the tangent to f at $x = a$ is the reciprocal of the slope of the tangent to g at $x = b$. This reciprocal relationship is often written:

$$\dfrac{dy}{dx} = \dfrac{1}{\frac{dx}{dy}} \text{ or } \dfrac{dx}{dy} = \dfrac{1}{\frac{dy}{dx}}$$

We can use this relationship to find the derivative of an inverse function without first finding the inverse of the function.

Example 17
If $f(x) = x^3 + x + 1$, find $(f^{-1})'(x)$ (that is, the derivative of the inverse of f.)

DERIVATIVES

Solution 1:

Interchanging x and y: $x = y^3 + y + 1$. Differentiating with respect to y:

$\frac{dx}{dy} = 3y^2 + 1$. Since $\frac{dy}{dx} = \frac{1}{\frac{dx}{dy}}$, we have $\frac{dy}{dx} = \frac{1}{3y^2 + 1}$.

Solution 2:

The solution can also be obtained using implicit differentiation. To find the inverse, interchange x and y to give $x = y^3 + y + 1$. Then use implicit differentiation: $1 = 3y^2 \left(\frac{dy}{dx}\right) + \left(\frac{dy}{dx}\right)$. Solve for $\frac{dy}{dx}$: $1 = (y^2 + 1)\left(\frac{dy}{dx}\right)$ or

$\frac{1}{3y^2 + 1} = \frac{dy}{dx}$, which is the same answer obtained using the reciprocal relationship in Solution 1.

Note that in addition to the given function being differentiable and defined so that its inverse exists (i.e., it must be one-to-one), we must also satisfy $f'(x) \neq 0$ to prevent the possibility of dividing by 0. Note that in the previous example $f'(x) = 3x^2 + 1$ is always greater than 0, and that f is both differentiable and one-to-one.

PARAMETRIC FUNCTIONS

While a rectangular equation in two variables provides information about the relationship between x and y, parametric equations provide information about how x and y change with respect to a third variable, usually t or θ. Graphs can be obtained using a table of values where t values are chosen and corresponding x and y values are found. Or, put your calculator in parametric mode (consult your user's manual), enter expressions for x and y, and graph.

Example 18
Sketch the curve represented by the parametric equations, using arrows to indicate the direction of the curve.

a. $x = 3t + 1$ and $y = 2t - 1$

b. $x = 2\cos\theta$ and $y = 4\sin\theta$ $\quad 0 \leq \theta \leq 2\pi$

PART II

Solutions:

a. Using a table of values and choosing t values, we have:

t	-2	-1	0	1	2
x	-5	-2	1	4	7
y	-5	-3	-1	1	3

or, by setting your calculator in parametric mode, the graph is:

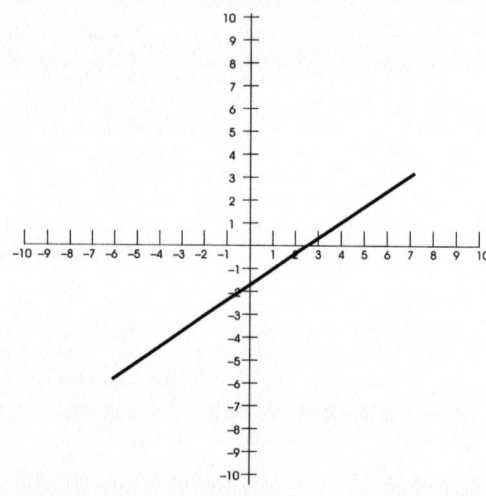

With the arrow added to indicate the direction of the curve, we have:

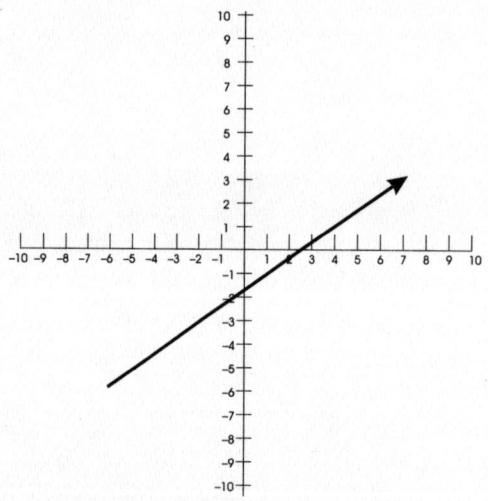

b. Here the parameter is θ, rather than t, but the principle is the same. Choosing θ values and finding x and y yields:

θ	0	$\pi/4$	$\pi/2$	$3\pi/4$	π
x	2	$\sqrt{2}$	0	$-\sqrt{2}$	-2
y	0	$2\sqrt{2}$	4	$2\sqrt{2}$	0

DERIVATIVES

If you need more values to sketch the graph, choose more values for θ. Or, use your graphing calculator in parametric mode. (The first graph is from the calculator; the second has the arrows to indicate the direction of the curve as increases from 0 to 2π. Note the different scales.)

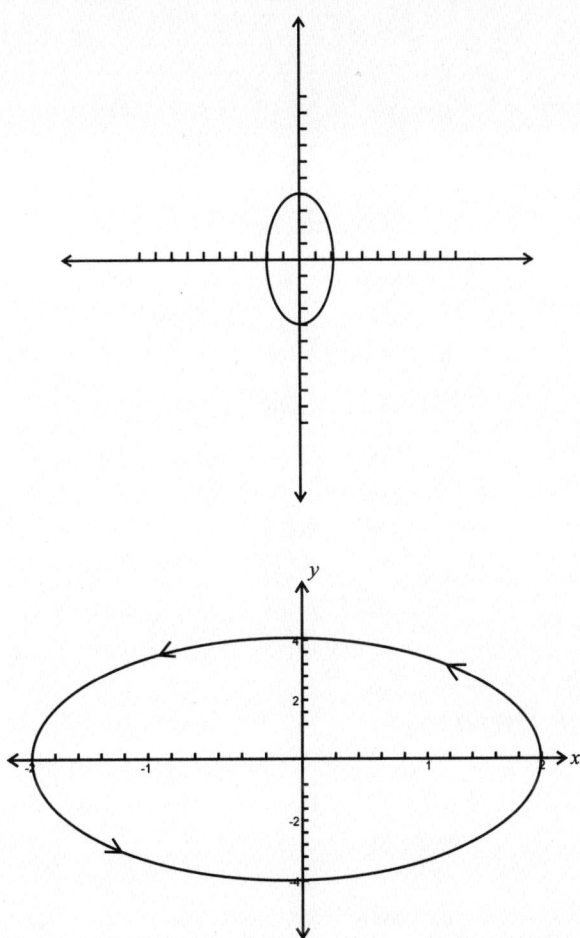

ELIMINATING THE PARAMETER

A set of parametric equations can be converted to rectangular form by eliminating the parameter. Solve one equation for the parameter and substitute into the other equation (this generally works for algebraic equations) or use trigonometric identities to eliminate the parameter for trigonometric equations.

Example 19
Write the rectangular equation by eliminating the parameter.

a. $x = 3t + 1$ and $y = 2t - 1$ b. $x = 2\cos\theta$ and $y = 4\sin\theta$ $0 \leq \theta \leq 2\pi$.

Solutions:

a. Since $x = 3t + 1$, $t = \dfrac{x-1}{3}$. Then substitute into the other equation:

$y = 2\left(\dfrac{x-1}{3}\right) - 1 = \dfrac{2}{3}x - \dfrac{5}{3}$. Note this is the equation of a line with slope $\dfrac{2}{3}$ and y intercept $-\dfrac{5}{3}$. Once the equation is converted to rectangular form, the idea of direction is lost, but the graph will match the first graph shown in Example 18a.

b. It is difficult to solve for θ, so solve for $\cos\theta$ and $\sin\theta$: $\cos\theta = \dfrac{x}{2}$ and $\sin\theta = \dfrac{y}{4}$. Now use $\sin^2\theta + \cos^2\theta = 1$ to write: $\left(\dfrac{x}{2}\right)^2 + \left(\dfrac{y}{4}\right)^2 = 1$ or $\dfrac{x^2}{4} + \dfrac{y^2}{16} = 1$. You should recognize this as the equation of an ellipse with center $(0, 0)$, vertices $(0, 4)$ and $(0, -4)$, minor axis of length 2 and note that this is the graph shown in Example 18b.

Parametric Form of the Derivative

When a plane curve is defined by parametric equations, derivatives can be found using:

$$\dfrac{dy}{dx} = \dfrac{\frac{dy}{dt}}{\frac{dx}{dt}}, \quad \dfrac{dx}{dt} \neq 0$$

$$\dfrac{d^2y}{dx^2} = \dfrac{\frac{d}{dt}\left[\frac{dy}{dx}\right]}{\frac{dx}{dt}}$$

$$\dfrac{d^3y}{dx^3} = \dfrac{\frac{d}{dt}\left[\frac{d^2y}{dx^2}\right]}{\frac{dx}{dt}}$$

The pattern continues in this fashion.

DERIVATIVES

Example 20

Find $\dfrac{dy}{dx}$ and $\dfrac{d^2y}{dx^2}$ for the parametric equations $x = 3t + 1$ and $y = 2t - 1$.

Solution:

$$\frac{dx}{dy} = \frac{\frac{d}{dt}(2t-1)}{\frac{d}{dt}(3t+1)} = \frac{2}{3} \quad \text{and} \quad \frac{d^2y}{dx^2} = \frac{\frac{d}{dt}\left[\frac{2}{3}\right]}{3} = 0$$

Example 21

Find $\dfrac{dy}{dx}$ and $\dfrac{d^2y}{dx^2}$ for the parametric equations $x = 2\cos\theta$ and $y = 4\sin\theta$.

Solution:

$$\frac{dy}{dx} = \frac{\frac{d}{d\theta}[4\sin\theta]}{\frac{d}{d\theta}[2\cos\theta]} = \frac{4\cos\theta}{-2\sin\theta} = -2\cot\theta \quad \text{and}$$

$$\frac{d^2y}{dx^2} = \frac{\frac{d}{d\theta}[-2\cot\theta]}{-2\sin\theta} = \frac{-2(-\csc^2\theta)}{-2\sin\theta} = -\csc^3\theta$$

POLAR FUNCTIONS

The polar coordinate system provides another way of graphing curves. A point in the polar coordinate system consists of an ordered pair (r, θ), where r is the directed distance from the pole to the point, and θ is the angle from the pole to the line segment containing the point.

Example 22
Plot the following points in a polar coordinate system.

$$A\left(2, \frac{\pi}{4}\right) \quad B\left(-1, \frac{\pi}{2}\right) \quad C\left(3, -\frac{\pi}{4}\right) \quad D\left(-2, -\frac{3\pi}{4}\right).$$

Solution:

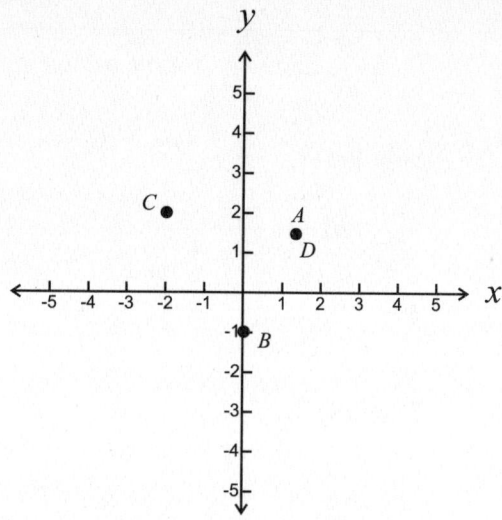

A positive angle is measured counterclockwise, and a positive r value is measured from the pole on the terminal side of the angle. A negative r value is found by measuring along the extension of the terminal side of the angle *through* the pole. A negative angle is measured clockwise. Notice that points A and D are the same. While each point in the rectangular plane is represented by a unique pair of coordinates, each point in the polar coordinate plane has an infinite number of ordered pair representations.

The following formulas are used to convert from polar to rectangular coordinates and vice versa:

From Polar to Rectangular Coordinates
$x = r \cos \theta$
$y = r \sin \theta$

From Rectangular to Polar Coordinates
$r^2 = x^2 + y^2$

$\tan \theta = \dfrac{y}{x}$

Example 23
Convert from polar to rectangular coordinates:

a. $\left(-1, \dfrac{\pi}{2}\right)$ b. $\left(2, \dfrac{\pi}{4}\right)$

Solutions:

a. Since $r = -1$ and $\theta = \dfrac{\pi}{2}$, $x = -1\cos\dfrac{\pi}{2} = -1(0) = 0$ and

DERIVATIVES

$y = -1\sin\frac{\pi}{2} = -1(1) = -1$. The rectangular coordinates are $(0, -1)$.

b. $r = 2$, $\theta = \frac{\pi}{4}$ so $x = 2\cos\frac{\pi}{4} = 2\left(\frac{\sqrt{2}}{2}\right) = \sqrt{2}$ and

$y = 2\sin\frac{\pi}{4} = 2\left(\frac{\sqrt{2}}{2}\right) = \sqrt{2}$. The rectangular coordinates are

$(\sqrt{2}, \sqrt{2})$.

Example 24
Find two sets of polar coordinates for the point $(-1, 1)$ for $0 \leq \theta \leq 2\pi$.

Solution:

Since $x = -1$ and $y = 1$, $r^2 = (-1)^2 + (1)^2 = 2$ and $r = \pm\sqrt{2}$. (Remember to use $\pm$ when taking the square root of both sides of the equation.) $\tan\theta = \frac{1}{-1}$ so the reference angle is $\frac{\pi}{4}$. The point $(-1, 1)$ is in quadrant II, so pair r and θ appropriately to give solutions $\left(\sqrt{2}, \frac{3\pi}{4}\right)$ and $\left(-\sqrt{2}, \frac{7\pi}{4}\right)$.

GRAPHING POLAR EQUATIONS

Polar equations can be graphed by plotting points, using a graphing calculator in polar function mode, or by first converting to rectangular form.

Example 25

Graph (a) $r = 3$ (b) $\theta = \frac{\pi}{4}$.

Solutions:

a. Since $x^2 + y^2 = r^2$, this equation converts to $x^2 + y^2 = 9$, the equation of a circle with center $(0, 0)$ and radius 3. Using a graphing calculator in polar mode gives:

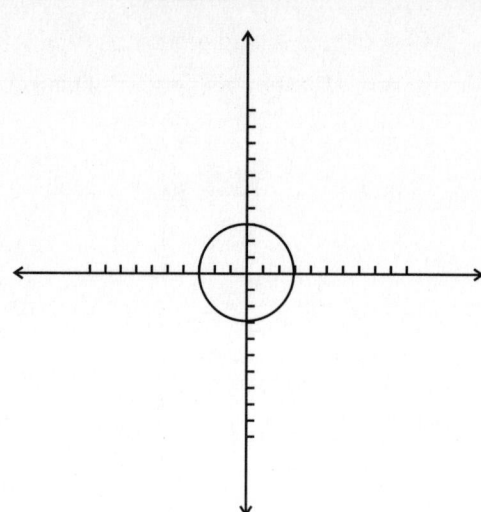

b. Converting the equation to $\tan\left(\dfrac{\pi}{4}\right) = \dfrac{y}{x}$ gives $\dfrac{y}{x} = 1$ or $y = x$, the equation of a line. Note that this equation cannot be graphed with the calculator in polar mode since it cannot be solved for r. However, we can reason that r takes on all values as θ remains $\dfrac{\pi}{4}$, producing the graph of a line. Or, put your calculator back in rectangular mode and graph $y = x$:

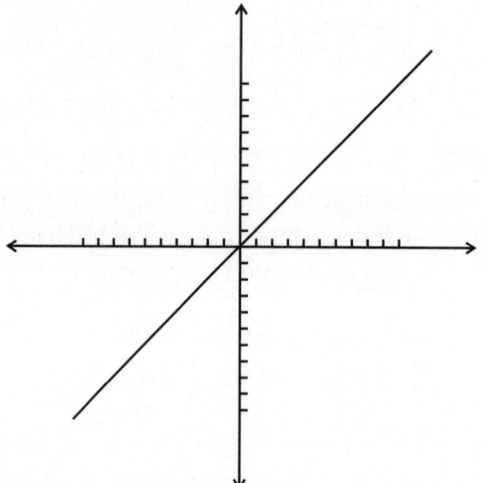

Example 26
Graph $r = 4\sin 2\theta$

Solution:

Using a table of values and choosing special angles and quadrantal angles for θ:

θ	0	$\pi/6$	$\pi/4$	$\pi/3$	$\pi/2$
r	0	$2\sqrt{3}$	4	$2\sqrt{3}$	0

DERIVATIVES

(Extend the table until you have sufficient points to sketch the curve.) Or, using a graphing calculator, you obtain:

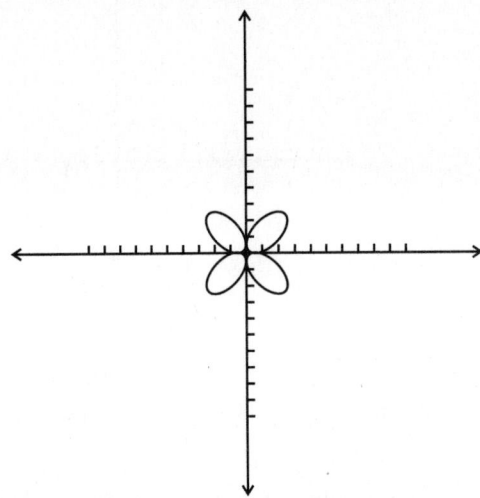

DERIVATIVES OF A FUNCTION IN POLAR FORM

To find the first derivative of a function in the form $r = f(\theta)$ at the point (r, θ),

$$\frac{dy}{dx} = \frac{f(\theta)\cos\theta + f'(\theta)\sin\theta}{-f(\theta)\sin\theta + f'(\theta)\cos\theta}$$

Example 27

Find $\frac{dy}{dx}$ if $r = 3 - 3\sin\theta$ at the point $\left(6, \frac{3\pi}{2}\right)$.

Solution:

$f(\theta) = 3 - 3\sin\theta$, $f'(\theta) = -3\cos\theta$ and $f\left(\frac{3\pi}{2}\right) = 3 - 3\sin\frac{3\pi}{2} = 6$,

$f'(\theta) = -3\cos\frac{3\pi}{2} = 0$ so

$$\frac{dy}{dx} = \frac{6\cos\frac{3\pi}{2} + 0\cdot\sin\frac{3\pi}{2}}{-6\sin\frac{3\pi}{2} + 0\cdot\cos\frac{3\pi}{2}} = \frac{6(0) + 0}{-6(-1) + 0} = \frac{0}{6} = 0$$

Notice that the line through the point $\left(6, \frac{3\pi}{2}\right)$ that has a slope of 0 is a horizontal line.

PART II

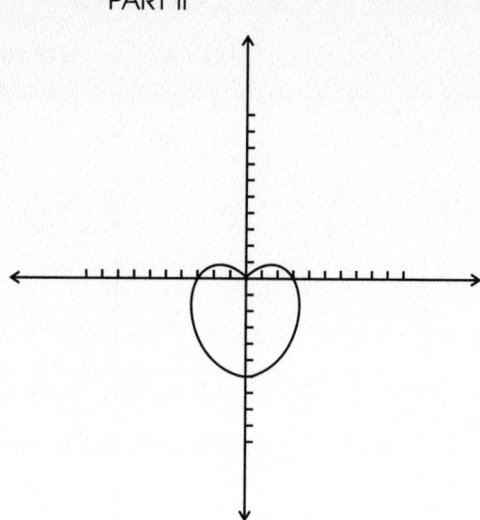

Vector-Valued Functions

A vector-valued function can be written as $r(t) = f(t)\mathbf{i} + g(t)\mathbf{j}$ where $f(t)$ and $g(t)$ are real-valued functions of the parameter t. We can also write $r(t) = (f(t), g(t))$. A vector-valued function can be graphed by plotting points from a table where t values are chosen and points $(f(t), g(t))$ are plotted, or by setting a graphing calculator in parametric mode and using $x(t) = f(t)$ and $y(t) = g(t)$.

Example 28
Sketch the plane curve represented by $r(t) = 2t\mathbf{i} + (3t + 1)\mathbf{j}$, $0 \le t \le 3$.

Solution:
Using a table of values:

t	0	1	2	3
$f(t)$	0	2	4	6
$g(t)$	1	4	7	10

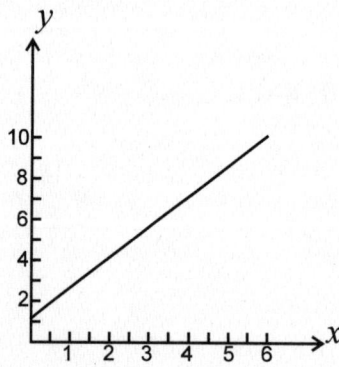

or, setting your graphing calculator in parametric mode and letting $x(t) = 2t$ and $y(t) = 3t + 1$ and tMin $= 0$, tMax $= 3$ with tStep $= .13$ gives:

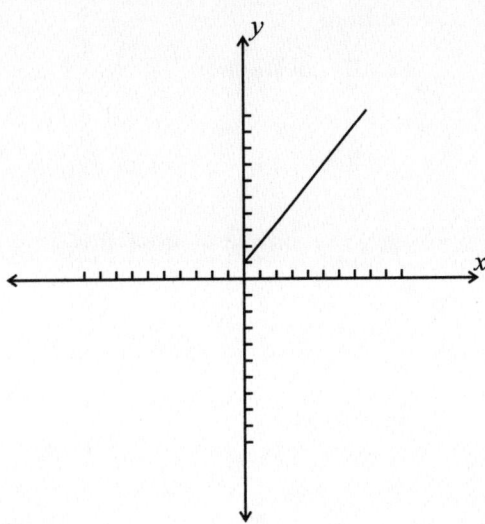

Domain of a Vector-Valued Function

The domain of a vector-valued function $r(t) = f(t)i + g(t)j$ is the intersection of the domains of $f(t)$ and $g(t)$, unless stated otherwise.

Example 29
Find the domain of each vector-valued function:

a. $r(t) = \sqrt{2t}\,i + e^t j$
b. $r(t) = \dfrac{1}{t}i + \ln t\,j$

Solutions:

a. The domain of $f(t) = \sqrt{2t}$ is $(-\infty, \infty)$, and the domain of $g(t) = e^t$ is also $(-\infty, \infty)$. Therefore the domain of $r(t)$ is $(-\infty, \infty)$.

b. The domain of $f(t) = \dfrac{1}{t}$ is $(-\infty, 0) \cup (0, \infty)$. The domain of $g(t) = \ln t$ is $(0, \infty)$. The intersection of these is the domain of $r(t)$ and equals $(0, \infty)$.

Limits and Continuity of Vector-Valued Functions

By considering the components of $r(t)$ individually, that is, $f(t)$ and $g(t)$, we can apply our previous rules for finding limits. Similarly, $r(t)$ will be continuous at $t = a$ if both $f(t)$ and $g(t)$ are continuous at $t = a$.

Example 30
Evaluate each limit.

a. $\lim\limits_{t \to 0}(ti + e^t j)$
b. $\lim\limits_{t \to 0}\left(\dfrac{\sin t}{t}i + \dfrac{1-\cos t}{t}j\right)$

Solutions:

a. By substitution, we have $\lim_{t \to 0}(ti + e^t j) = 0 \cdot i + e^0 j = j$

b. Recall $\lim_{t \to 0} \dfrac{\sin t}{t} = 1$ and $\lim_{t \to 0} \dfrac{1-\cos t}{t} = 0$ so that

$$\lim_{t \to 0}\left(\frac{\sin t}{t}i + \frac{1-\cos t}{t}j\right) = 1 \cdot i + 0 \cdot j = 1$$

Example 31
Determine the interval(s) on which each vector-valued function is continuous.

a. $r(t) = 2ti + (3t + 1)j$ b. $r(t) = \sqrt{3-t}\,i + \dfrac{1}{t-2}j$

Solutions:

a. The domain of $r(t)$ is $(-\infty, \infty)$ and r is continuous on $(-\infty, \infty)$.

b. Examining the components of the vector-valued function, the domain of $f(t) = \sqrt{3-t}$ is $(-\infty, 3)$, the domain of $g(t) = \dfrac{1}{t-2}$ is $(-\infty, 2) \cup (2, \infty)$. Therefore the domain of $r(t)$ is $(-\infty, 2) \cup (2, 3]$. The vector-valued function will be continuous on its domain. Note $r(t)$ is *not* continuous at $t = 2$, the vertical asymptote of $g(t)$.

THE DERIVATIVE OF A VECTOR-VALUED FUNCTION

If $r(t) = f(t)i + g(t)j$, then $r'(t) = f'(t)i + g'(t)j$, which means we can continue to work with the derivative of the components of r to find $r'(t)$.

Example 32
Find $r'(t)$ and $r''(t)$ if $r(t) = t^3 i - (4t + 1)^2 j$

Solution:
$r'(t) = 3t^2 i - 2(4t + 1)(4)j = 3t^2 i - 8(4t + 1)j$ where the derivative of $g(t) = (4t + 1)^2$ is found using the chain rule.
$r''(t) = 6ti - 8(4)j = 6ti - 32j$.

Example 33
Find $r'(t)$ and $r''(t)$ if $r(t) = e^{2t}i + \ln(t^2 + 1)j$.

DERIVATIVES

Solutions:

$$r'(t) = 2e^{2t}i + \frac{2t}{t^2+1}j \text{ and } r''(t) = 4e^{2t}i + \left(\frac{-2t^2+2}{(t^2+1)^2}\right)j$$

DERIVATIVE AT A POINT

The slope of a tangent line at a point on a curve and instantaneous velocity can now be computed easily using the differentiation rules rather than the limit definitions.

Example 34
Find the equation of the tangent line at the indicated point. Write each answer in slope-intercept form.

a. $f(x) = x^3 + 2x + 1$ at $(1, 4)$

b. $g(x) = (2x - 5)^3$ at $(3, 1)$

c. $h(x) = \sin 6x$ at $\left(\frac{\pi}{4}, -1\right)$

Solutions:

a. Since $f'(x) = 3x^2 + 2,\ f'(1) = 3(1)^2 + 2 = 5,$ and the slope of the tangent at the given point is $m = 5$. Then $y - y_1 = m(x - x_1),\ y - 4 = 5(x - 1), y = 5x - 1$. You can check the reasonableness of your answer by graphing the given function and the tangent line in the same viewing window.

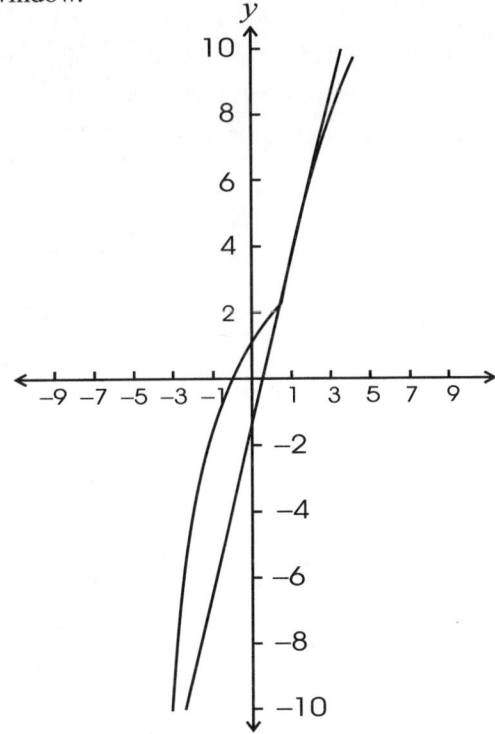

b. $g'(x) = 3(2x-5)^2 (2)$ using the chain rule. Then $g'(3) = 3(1)^2 \cdot 2 = 6$. $y - 1 = 6(x - 3)$, $y = 6x - 17$. A quick sketch with the graphing calculator will help you detect any errors:

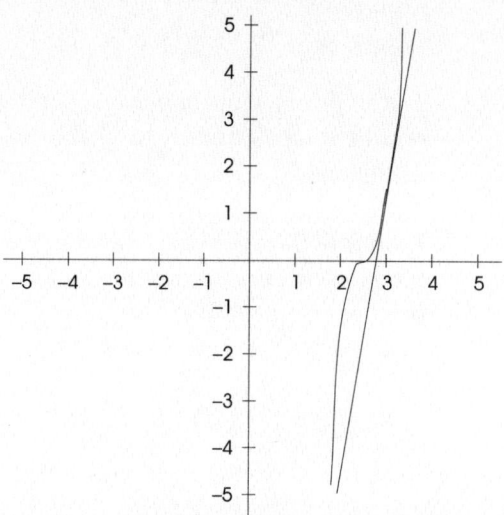

c. $h'(x) = 6 \cos 6x$ and $h'\left(\dfrac{\pi}{4}\right) = 6\cos 6\left(\dfrac{\pi}{4}\right) = 6(0) = 0$. Since any horizontal line has a slope of 0, we know this line must be a horizontal line with equation $y = -1$. The graph shown below has a viewing window $-2\pi \leq x \leq 2\pi$ with a scale of $\dfrac{\pi}{4}$, $-3 \leq y \leq 3$ with a scale of 1:

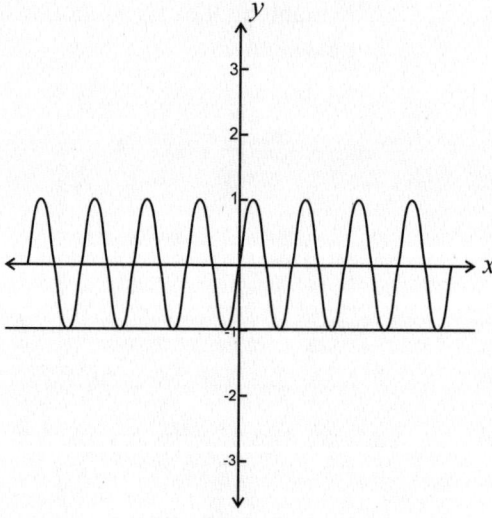

DERIVATIVES

Example 35

Find the equation of the tangent line to $x^2 + 4y^2 = 100$ at $(-8, 3)$.

Solution:

The slope of the tangent must be found using implicit differentiation:

$2x + 8yy' = 0$ so $y' = -\dfrac{2x}{8y} = -\dfrac{x}{4y}$. Find y' at the point $(-8, 3)$:

$y' = -\dfrac{(-8)}{4(3)} = \dfrac{2}{3}$. Now, find the equation of the tangent line:

$y - 3 = \dfrac{2}{3}(x + 8)$ or $y = \dfrac{2}{3}x + \dfrac{25}{3}$.

Example 36

Find the equation of the tangent line and the normal line to $f(x) = e^{2x}$ at $(0, 1)$.

Solution:

$f'(x) = 2e^{2x}$ and $f'(0) = 2e^{2(0)} = 2$. Then $y - 1 = 2(x - 0)$ or $y = 2x + 1$. The normal line is perpendicular to the tangent line, so it has slope $m = -\dfrac{1}{2}$. The equation of the normal line is $y - 1 = -\dfrac{1}{2}(x - 0)$ or $y = -\dfrac{1}{2}x + 1$. Using a graphing calculator to graph the function, the tangent line, and the normal line in a standard viewing window gives:

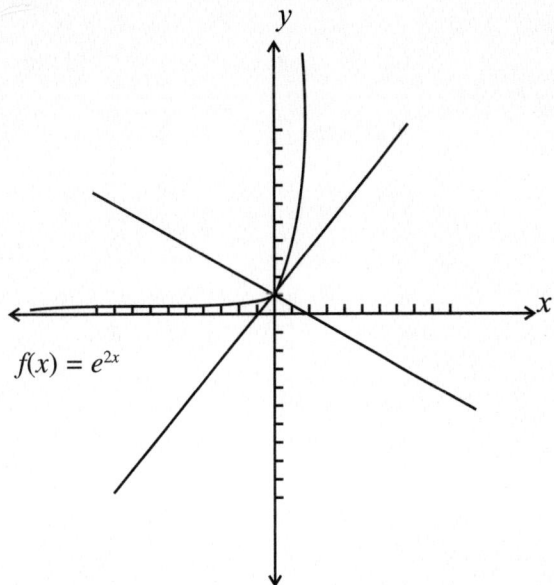

$f(x) = e^{2x}$

The tangent line and normal line do not appear to be perpendicular, but using a function, such as ZSQR or a similar function (check the calculator manual), will square the viewing window and help you view the graph:

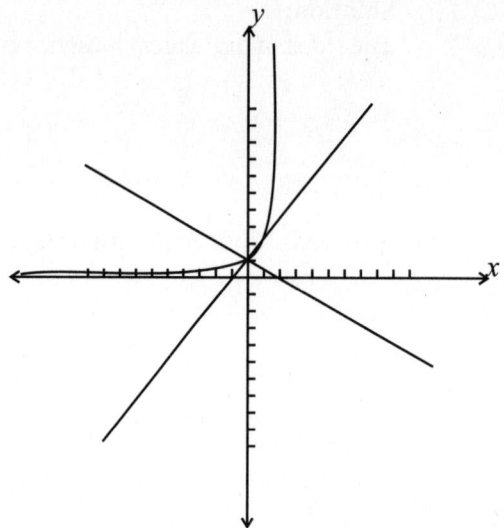

Example 37
Find the equation of any horizontal tangents to $f(x) = x^2 + 2x + 3$.

Solution:
Horizontal lines have $m = 0$. Thus we should solve the equation $f'(x) = 0$. Here $f'(x) = 2x - 2$ and $2x - 2 = 0$ when $x = 1$. When $x = 1$, $f(1) = (1)^2 - 2(1) + 3 = 2$. Therefore $y = 2$ is the equation of the horizontal tangent.

Example 38

Find the equation of any vertical tangents to $f(x) = x^{\frac{2}{3}} + 2$.

Solution:

$f'(x) = \frac{2}{3} x^{-\frac{1}{3}} = \frac{2}{3\sqrt[3]{x}}$. The derivative is undefined when $x = 0$, which

means the slope of the tangent line at $x = 0$ is undefined and hence is a vertical tangent line. The equation of the tangent line is $x = 0$.

Example 39
Discuss the tangent line to $f(x) = |x|$ at $x = 0$.

DERIVATIVES

Solution:

$\lim_{x \to 0^+} \frac{|x|-0}{x-0} = 1$ and $\lim_{x \to 0^-} \frac{|x|-0}{x-0} = -1$ so the function is not differentiable at $x = 0$. The graph does not have a tangent line at $x = 0$. In general, a graph does not have a tangent line at a sharp point.

RELATED RATES

The derivative can be viewed as the rate of change of y over x, and velocity (also the derivative) as the rate of change of distance over time. Other rates of change involve change over time and will require implicit differentiation to compute the derivatives.

Example 40
Differentiate each equation with respect to t.

a. $A = \pi r^2$ b. $V = \frac{4}{3}\pi r^3$ c. $c^2 = a^2 + b^2$ d. $\tan \theta = \frac{y}{50}$

Solutions:

a. Take the derivative of each side of the equation with respect to t:

$$\frac{dA}{dt} = 2\pi r \frac{dr}{dt}.$$

b. $\frac{dV}{dt} = \frac{4}{3}(3)\pi r^2 \frac{dr}{dt}$ and simplifying, $\frac{dV}{dt} = 4\pi r^2 \frac{dr}{dt}$

c. $2c\frac{dc}{dt} = 2a\frac{da}{dt} + 2b\frac{db}{dt}$ and simplifying, $c\frac{dc}{dt} = a\frac{da}{dt} + b\frac{db}{ct}$.

d. $\sec^2 \theta \frac{d\theta}{dt} = \frac{1}{50}\frac{dy}{dt}$.

The following guidelines may help you solve related rates problems:

1. Find an equation that describes the relationship between the variables. This may be an equation from your past, such as the Pythagorean Theorem, or one given in the problem.

2. Express the given information symbolically.

3. Differentiate both sides of the equation with respect to time.

4. Use information given in the problem to substitute into the equation, and solve for whatever is required.

Example 41
Oil spills into a lake in a circular pattern. If the radius of the circle increases at a constant rate of 3 feet per second, how fast is the area of the spill increasing at the end of 30 minutes?

Solution:

Use the formula for the area of a circle to relate the variables, $A = \pi r^2$.

$\frac{dr}{dt} = 3$ ft/sec since the radius increases at this rate. From Example 13,

$\frac{dA}{dt} = 2\pi r \frac{dr}{dt}$. Substitute into this equation the value for $\frac{dr}{dt}$ and $r = 5{,}400$ feet (this is computed from the fact that in 30 minutes, which equals 1,800 seconds, the radius will be 3 × 1,800 feet = 5,400 feet).

$\frac{dA}{dt} = 2\pi(5{,}400 \text{ ft})(3 \text{ ft/sec}) = 32{,}400\pi \text{ ft}^2/\text{sec}$.

Example 42
Air is being pumped into a spherical balloon at a rate of 6 cubic inches per minute. Find the rate of change of the radius when the radius is 1.5 inches.

Solution:

The formula for the volume of a sphere is $V = \frac{4}{3}\pi r^3$. You are given

$\frac{dV}{dt} = 6$ in³/min as the rate of change of volume with respect to time. You are to find $\frac{dr}{dt}$ when $r = 1.5$ in. Substitute into the derivative found in

Example 13: $\frac{dV}{dt} = 4\pi r^2 \frac{dr}{dt}$ so that $6 \text{ in}^3/\text{min} = 4\pi(1.5 \text{ in})^2 \left(\frac{dr}{dt}\right)$ and

solving for $\frac{dr}{dt}$ gives $\frac{dr}{dt} = \frac{2}{3\pi}$ in /min.

Example 43
A 13-foot ladder is leaning against the wall of a house. The base of the ladder slides away from the wall at a rate of 0.75 feet per second. How fast is the top of the ladder moving down the wall when the base is 12 feet from the wall?

DERIVATIVES

Solution:
A sketch may help:

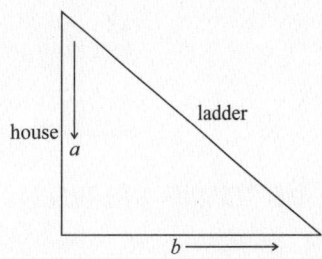

Using the Pythagorean Theorem: $13^2 = a^2 + b^2$, where the length of the ladder is 13 feet. We are also given that $\frac{db}{dt} = 0.75$ ft/sec, which indicates that the ladder slides away from the wall at this rate. Substituting into the derivative found in Example 13, $c\frac{dc}{dt} = a\frac{da}{dt} + b\frac{db}{dt}$, and recognizing that since the length of the ladder does not change, $\frac{dc}{dt} = 0$, we have:

$0 = a\left(\frac{da}{dt}\right) + b(0.75 \text{ ft/sec})$ and we know $a = 5$ when $b = 12$ (use the Pythagorean Theorem, or recall the 5-12-13 right triangle) and we have:

$0 = (5 \text{ ft})\left(\frac{da}{dt}\right) + (12 \text{ ft})(0.75 \text{ ft/sec})$ or $\frac{da}{dr} = -\frac{5}{9}$ ft/sec. Note that the answer is negative to represent the ladder sliding down the wall.

Example 44
A weather balloon is released 50 feet from an observer. It rises at a rate of 8 feet per second. How fast is the angle of elevation changing when the balloon is 50 feet high?

Solution:
Draw a quick sketch:

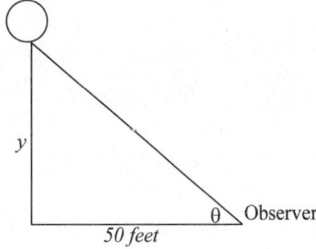

From the definition of the tangent of an angle, we have: $\tan\theta = \frac{y}{50}$, and we are given that $\frac{dy}{dt} = 8$ ft/sec since the balloon is rising at that rate. Using the derivative found in Example 13, $\sec^2\theta \frac{d\theta}{dt} = \frac{1}{50}\frac{dy}{dt}$, we can substitute for $\frac{dy}{dt}$, but not for $\sec^2\theta$. However, using the fact that when the balloon is 50 feet high, we have $\tan\theta = \frac{50}{50}$, which means $\theta = \frac{\pi}{4}$. Now complete the substitutions: $\sec^2\left(\frac{\pi}{4}\right)\left(\frac{d\theta}{dt}\right) = \frac{1}{50}(8 \text{ ft /sec})$ and $\frac{d\theta}{dt} = \frac{2}{25}$ rad /sec.

INCREASING AND DECREASING INTERVALS

A function is increasing when y values increase as x values increase; it is decreasing when y values decrease as x values increase. A graphing calculator provides a general idea of where a function increases or decreases, but examining the sign changes of the first derivative produce exact information about where these changes occur. To determine intervals where a function is increasing or decreasing:

1. Find critical numbers, that is, x values in the domain of the function where the first derivative equals 0 and where the first derivative is undefined.

2. Draw a number line, and draw vertical lines at each critical number found in step 1.

3. Choose a number in each region and record whether the first derivative is positive or negative in the region.

4. Intervals where the first derivative is greater than 0 are intervals where the function is increasing; intervals where the first derivative is less than 0 are intervals where the function is decreasing.

5. Check answers with a graph.

Note: Steps 2, 3, and 4 can be done by graphing the first derivative and intervals where the derivative is above the x axis correspond to intervals where the original function is increasing, and intervals where the derivative is below the x axis correspond to intervals where the original function is decreasing. Example 45 is done both ways.

Example 45
Find the intervals on which $f(x) = x^3 - 4x^2 + 1$ is increasing or decreasing.

Solution:
To find the critical numbers, set the derivative equal to 0, and determine any values where the first derivative is undefined: $f'(x) = 3x^2 - 8x$ and $3x^2 - 8x = 0$ when $x = 0$ and $x = \frac{8}{3}$. The derivative is defined everywhere (note the first

derivative is a polynomial). Use a number line with the critical numbers marked, and determine whether the derivative is positive or negative in each region:

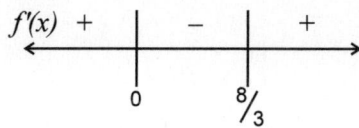

Using the note above, examine the graph of the derivative in a standard viewing window:

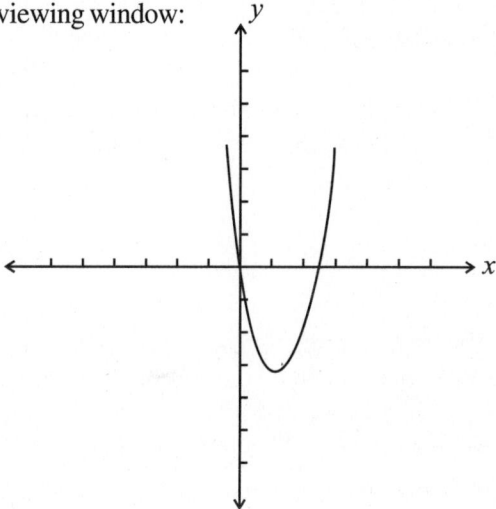

We see the same information as was obtained using the number line. Therefore, the function is increasing on $(-\infty, 0)$ and $\left(\frac{8}{3}, \infty\right)$ and decreasing on $\left(0, \frac{8}{3}\right)$.

Example 46
Find the intervals on which $f(x) = \dfrac{x^2}{x^2 - 4}$ is increasing and decreasing.

Solution:

Finding the derivative using the quotient rule gives

$$f'(x) = \frac{(x^2 - 4)(2x) - (x^2)(2x)}{(x^2 - 4)^2} = \frac{-8x}{(x^2 - 4)^2}.$$ Setting the numerator equal to 0 gives critical numbers where the derivative equals 0 (at $x = 0$ here), and setting the denominator equal to 0 gives critical numbers where the derivative is undefined (at $x = -2$ and $x = 2$). Note that -2 and 2 are *not* critical numbers because they are not in the domain of the original function. However, the graph can change from increasing to decreasing and vice versa around these values, so they should be used on your number line graph as split points.

PART II

Use either the number line approach or the graph of the first derivative (shown below) to conclude that the function is increasing on $(-\infty, -2)$ and $(-2, 0)$ and decreasing on $(0, 2)$ and $(2, \infty)$.

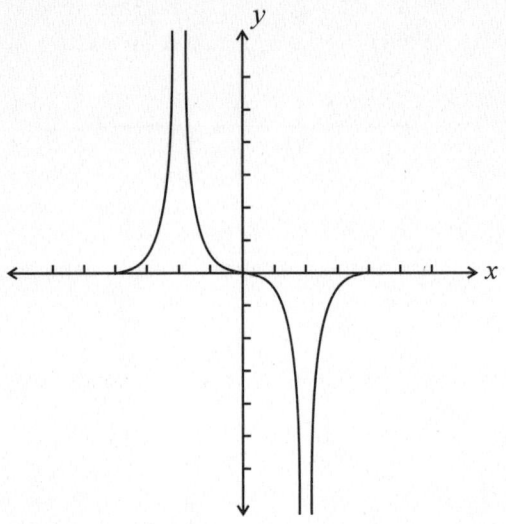

Extrema

Extrema of a function are the maximum and/or minimum y values of the function (also called the absolute maximum and absolute minimum). The relative extrema of a function are the maximum or minimum value(s) within an open interval. The following statements are used to verify the existence of extrema and to locate extrema:

1. A continuous function on a closed interval $[a, b]$ has both a maximum and a minimum (The Extreme Value Theorem).

2. If a function has a relative maximum or relative minimum, it must occur at a critical number.

Extrema on a Closed Interval

If the function is continuous on a closed interval $[a, b]$, use the following steps to find the absolute maximum and absolute minimum guaranteed by the Extreme Value Theorem:

1. Find critical numbers.

2. Evaluate the function at each critical number.

3. Evaluate $f(a)$ and $f(b)$.

4. The largest value from steps 2 and 3 is the absolute maximum, and the smallest value is the absolute minimum.

DERIVATIVES

5. Use a graph to check the accuracy of your answers (remember to set an appropriate viewing window).

Note that the absolute maximum or minimum can occur at an endpoint of the interval or at a critical number, and that the y value is the maximum or minimum.

Example 47
Find the absolute maximum and minimum for $f(x) = x^3 - 4x^2 + 1$ on $[-1, 5]$.

Solution:

The critical numbers are $x = 0$ and $x = \frac{8}{3}$ (found in Example 18). Evaluating f at each critical number gives:

$$f(0) = 0^3 - 4(0)^2 + 1 = 1; \quad f\left(\frac{8}{3}\right) = \left(\frac{8}{3}\right)^3 - 4\left(\frac{8}{3}\right)^2 + 1 = -\frac{229}{27} \approx -8.5$$

Evaluating f at the endpoints of the interval:

$$f(-1) = (-1)^3 - 4(-1)^2 + 1 = -4; \quad f(5) = (5)^3 - 4(5)^2 + 1 = 26$$

The absolute maximum of 26 occurs when $x = 5$ and the absolute minimum of $-\frac{229}{27}$ or -8.5 (if an approximation is allowed), occurs when $x = \frac{8}{3}$. Check your work with a graph (window: $-1 \leq x \leq 5$, Scl 1, $-10 \leq y \leq 30$, Scl 5):

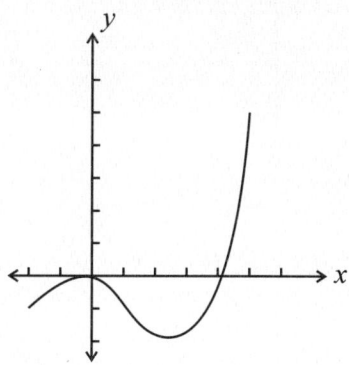

Example 48
Find the absolute maximum and minimum for $f(x) = \sin x$ on $[0, \pi]$.

Solution:

The function $f(x) = \sin x$ is continuous, and, hence, both an absolute maximum and absolute minimum must exist. Find the first derivative $f'(x) = \cos x$, set it equal to 0, and solve to find the critical number, $x = \dfrac{\pi}{2}$. Evaluate f at the critical number and at the endpoints of the interval

$$f\left(\dfrac{\pi}{2}\right) = \sin\dfrac{\pi}{2} = 1; \quad f(0) = \sin 0 = 0; \quad f(\pi) = \sin \pi = 0.$$

Thus, the absolute maximum of 1 occurs at $x = \dfrac{\pi}{2}$ and the absolute minimum of 0 occurs at $x = 0$ and $x = \pi$. A graph verifies our results (Window: $0 \leq x \leq \pi$, Scl $\dfrac{\pi}{4}$; Scl $-1 \leq y \leq 1$, Scl 0.5):

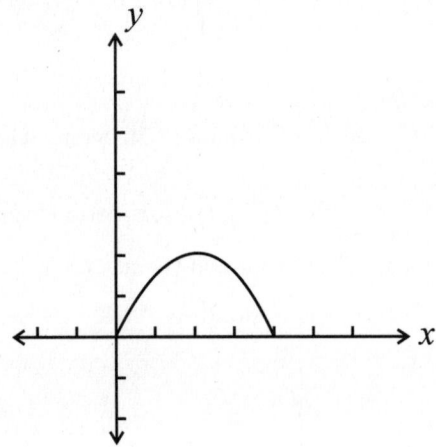

The Mean Value Theorem

The Mean Value Theorem states, if a function is continuous on a closed interval $[a, b]$ and differentiable on (a, b), then there exists c in (a, b) such that the slope of the tangent line at c equals the slope of the secant line through the endpoints of the closed interval, that is

$$f'(c) = \dfrac{f(b) - f(a)}{b - a}$$

Example 49

Find all values of c in the given interval that satisfy the conclusion of the Mean Value Theorem.

a. $f(x) = x^2$ on $[-1, 2]$ b. $f(x) = \sin\left(\dfrac{1}{2}x\right)$ on $\left[\dfrac{\pi}{2}, \dfrac{3\pi}{2}\right]$.

DERIVATIVES

Solution:

a. $f'(x) = 2x \cdot \dfrac{f(2)-f(-1)}{2-(-1)} = \dfrac{4-1}{3} = 1$. c must satisfy: $2c = 1$ or $c = \dfrac{1}{2}$.

The slope of the tangent line when $c = \dfrac{1}{2}$ is $f'\left(\dfrac{1}{2}\right) = 2\left(\dfrac{1}{2}\right) = 1$. To check this graphically, find the equation of the secant line. Since it passes through the points $(-1, 1)$ and $(2, 4)$, $y - 1 = 1(x + 1)$ or $y = x + 2$.

The equation of the tangent line at $x = \dfrac{1}{2}$ passes through the point $\left(\dfrac{1}{2}, \dfrac{1}{4}\right)$ and has slope 1: $y - \dfrac{1}{4} = 1\left(x - \dfrac{1}{2}\right)$ or $y = x - \dfrac{1}{4}$. Graph all three functions in the same viewing window. The graph below uses

$-3 \le x \le 5$ Scl = 1, $-10 \le y \le 10$, Scl = 1.

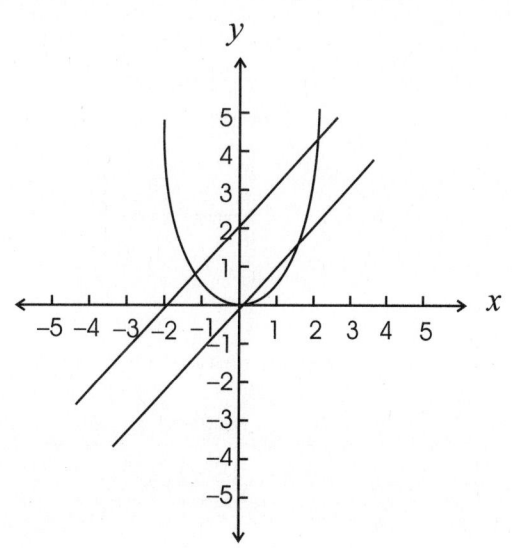

Notice that the secant line and tangent line have the same slope and are parallel.

b. $f'(x) = \dfrac{1}{2}\cos\left(\dfrac{1}{2}x\right)$.

$\dfrac{f\left(\dfrac{3\pi}{2}\right) - f\left(\dfrac{\pi}{2}\right)}{\dfrac{3\pi}{2} - \dfrac{\pi}{2}} = \dfrac{\sin\left(\dfrac{1}{2} \cdot \dfrac{3\pi}{2}\right) - \sin\left(\dfrac{1}{2} \cdot \dfrac{\pi}{2}\right)}{\pi} = \dfrac{\dfrac{\sqrt{2}}{2} - \dfrac{\sqrt{2}}{2}}{\pi} = 0$. Solving

$\dfrac{1}{2}\cos\left(\dfrac{1}{2}x\right) = 0$ gives $\dfrac{1}{2}x = \dfrac{\pi}{2}, \dfrac{3\pi}{2}$, so that $x = \pi, 3\pi, \ldots$ The c value in the interval $\left[\dfrac{\pi}{2}, \dfrac{3\pi}{2}\right]$ is $c = \pi$.

Relative Extrema

When a function is not continuous on a closed interval, we are not guaranteed that an absolute maximum and minimum exist. However, we can find relative extrema using the First Derivative Test. The test states that if the first derivative changes sign around a critical number c, then $f(c)$ is a relative maximum or minimum. If the signs change (from left to right) from positive to negative, $f(c)$ is a relative maximum; if the signs change (from left to right) from negative to positive, $f(c)$ is a relative minimum. If the signs don't change (from left to right), that critical value yields neither a relative maximum nor a relative minimum.

Example 50
Find the relative extrema for $f(x) = x^3 - 4x^2 + 1$.

Solution:

We have already found the critical numbers, $x = 0$ and $x = \frac{8}{3}$. We found the following sign changes in the first derivative:

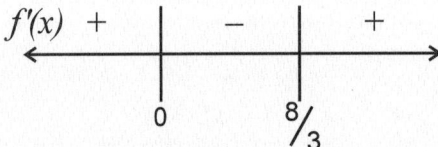

The First Derivative Test indicates that at $x = 0$ there is a relative maximum (sign changes from + to −) and that at $x = \frac{8}{3}$ there is a relative minimum (sign changes from − to +). To find the actual relative maximum and minimum, evaluate the original function at these values:

$$f(0) = 1 \quad \text{and} \quad f\left(\frac{8}{3}\right) = -\frac{229}{27}$$

Example 51
Find the open intervals on which $f(x) = 3x^4 - x^3 - 24x^2 + 12x$ is increasing or decreasing, and find all relative extrema.

Solution:

$f'(x) = 12x^3 - 3x^2 - 48x + 12$. To solve $f'(x) = 0$, use factoring (first common factoring, then factoring by grouping): $3(4x - 1)(x^2 - 4) = 0$ which means $x = \frac{1}{4}$, $x = -2$, or $x = 2$. Checking the sign of the first derivative around these critical numbers gives:

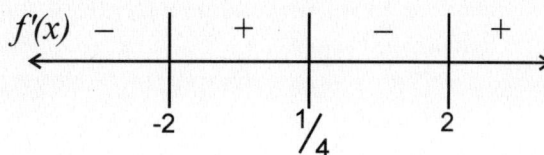

Thus, $f(x)$ is increasing on $\left(-2, \frac{1}{4}\right)$ and $(2, \infty)$, decreasing on $(-\infty, -2)$ and $\left(\frac{1}{4}, 2\right)$. $f(-2) = -64$ is a relative minimum (sign changes from $-$ to $+$).

$f\left(\frac{1}{4}\right) = \frac{383}{256}$ is a relative maximum (sign changes from $+$ to $-$). $f(2) = -32$ is a relative minimum (sign changes from $-$ to $+$). The graph of the original function and the graph of the derivative are shown below. Note that a standard viewing window will not provide a good graph in this case (our maximum and minimum are outside the boundaries of a standard viewing window), so you will have to adjust the window

$f(x)$ shown here with $-5 \leq x \leq 5$ Scl 1 and $-70 \leq y \leq 10$, Scl 10;

$f'(x)$ shown here with $-5 \leq x \leq 5$ Scl 1 and $-40 \leq y \leq 60$ Scl 10:

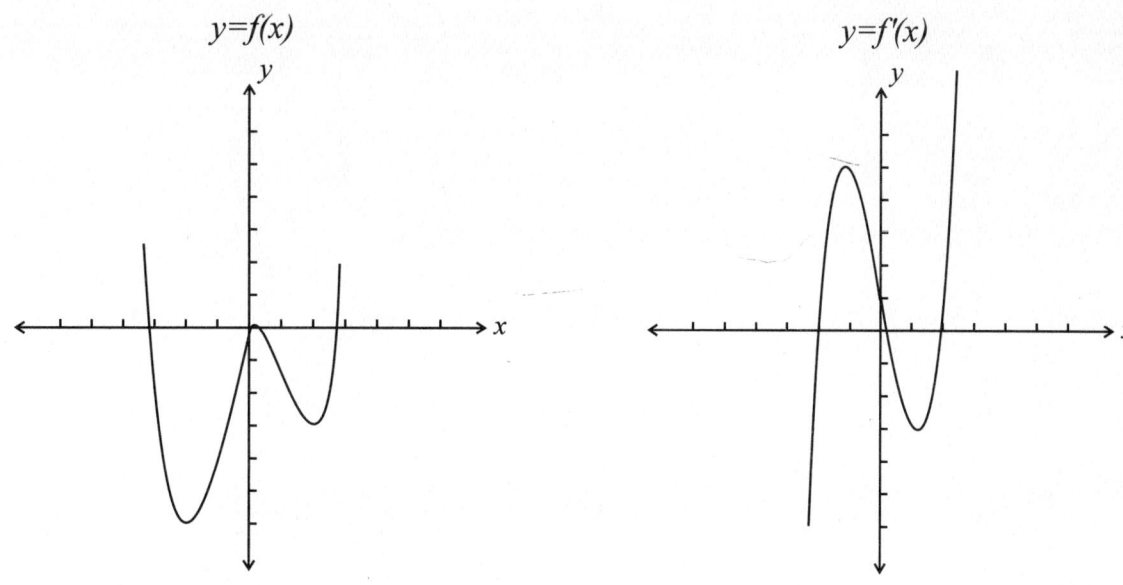

CONCAVITY

When the graph of a function is above its tangent lines, the graph is said to be concave up. When the graph of a function lies below its tangent lines, it is concave down. We can find intervals of concavity by examining the second derivative:

1. Find $f''(x)$.

2. Find x values where $f''(x) = 0$ and where $f''(x)$ is undefined.

3. Use a number line graph to determine the signs of $f''(x)$ around the values found in step 2 OR graph $y = f''(x)$.

4. The function is concave up where $f''(x) > 0$ or where the graph of $y = f''(x)$ is above the x axis. The function is concave down where $f''(x) < 0$ or where the graph of $y = f''(x)$ is below the x axis.

Example 52
Determine the intervals where $f(x) = x^3 - 4x^2 + 1$ is concave upward and downward.

Solution:

Find the second derivative: $f'(x) = 3x^2 - 8x$, $f''(x) = 6x - 8$. The second derivative equals 0 when $x = \dfrac{4}{3}$. Use a number line graph and check the signs of the second derivative around $\dfrac{4}{3}$:

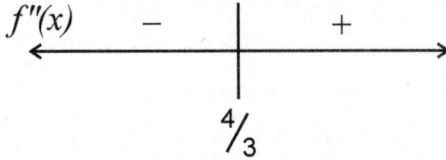

OR graph $y = f''(x)$:

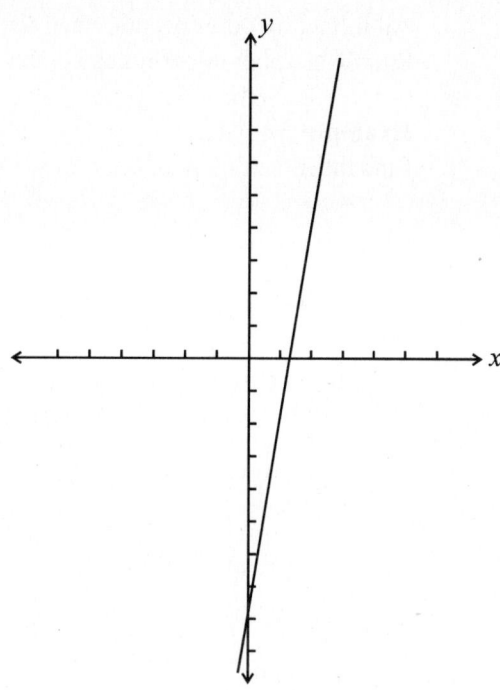

From either the number line graph or the graph of the second derivative, we find that is concave downward on $\left(-\infty, \dfrac{4}{3}\right)$ and concave upward on $\left(\dfrac{4}{3}, \infty\right)$.

Example 53

Determine the intervals where $f(x) = \dfrac{1}{x^2 - 4}$ is concave upward and concave downward.

Solution:
Find the second derivative: $f(x) = (x^2 - 4)^{-1}$; $f'(x) = -2x(x^2 - 4)^{-2}$ (using the chain rule); $f''(x) = -2(x^2 - 4)^{-3}(-3x^2 - 4)$ (using the product rule and the chain rule) and simplifying: $f''(x) = \dfrac{2(3x^2 + 4)}{(x^2 - 4)^3}$. $f''(x) = 0$ when the numerator is 0, but $3x^2 + 4 = 0$ has no real solutions. $f''(x)$ is undefined when the denominator equals 0, or when $x = \pm 2$. Using either a number line graph or the graph of $y = f''(x)$ reveal that $f''(x) > 0$ on $(-\infty, 2)$ and $(2, \infty)$ and so is concave upward there. $f''(x) < 0$ on $(-2, 2)$ and so is concave downward there.

Inflection Points

An inflection point occurs when the concavity changes. We can use the steps established above for finding concavity intervals to find inflection points.

Example 54
Find the inflection point(s), if any, for each function.

a. $f(x) = x^3 - 4x^2 + 1$ b. $f(x) = \dfrac{1}{x^2 - 4}$.

Solution:

a. From Example 25, we know the concavity changes when $x = \dfrac{4}{3}$.
Therefore, there is an inflection point at $\left(\dfrac{4}{3}, -\dfrac{101}{27}\right)$.

b. From Example 26, the concavity changes when $x = -2$. But, $x = -2$ is not a point on the graph of the function because -2 is not in the domain of the function. Therefore, there are no inflection points for the function.

Second Derivative Test

If we are interested in only whether a critical number is a maximum or minimum, we can use the Second Derivative Test. If $x = c$ is a critical number of $y = f(x)$ and $f''(c) > 0$ then $f(c)$ is a relative minimum, and if $f''(c) < 0$ then $f(c)$ is a relative maximum. If $f''(x) = 0$ or if $f''(c)$ is undefined, the Second Derivative Test fails, and you must return to the First Derivative Test to determine whether the critical number yields a maximum or minimum.

Example 55
Use the Second Derivative Test to find the relative extrema for each function.

a. $f(x) = 2x^4 + 3x^3 - 1$ b. $f(x) = \sin x + \cos x$ on $[0, 2\pi]$

Solution:

a. Find the second derivative: $f'(x) = 8x^3 + 9x^2$; $f''(x) = 24x^2 + 18x$.
Remember that critical numbers are found using the first derivative, so
$8x^3 + 9x^2 = 0$ when $x = 0$ and $x = -\dfrac{9}{8}$. Substitute each of these values into the second derivative:

$f''(0) = 0;$ $f''\left(-\dfrac{9}{8}\right) = \dfrac{81}{8} > 0$.

DERIVATIVES

Since $f''(0) = 0$, the Second Derivative Test fails. Returning to the First Derivative Test also gives no sign change at $x = 0$. Therefore, there is neither a maximum nor a minimum at $x = 0$.

Since $f''\left(-\dfrac{9}{8}\right) > 0$, there is a relative minimum when $x = -\dfrac{9}{8}$. The minimum is -2.08.

b. Find the first and second derivatives: $f'(x) = \cos x - \sin x$; $f''(x) = -\sin x - \cos x$. The critical numbers are at $x = \dfrac{\pi}{4}, \dfrac{5\pi}{4}$. Evaluate the second derivative at each critical number:

$$f''\left(\dfrac{\pi}{4}\right) = -\sqrt{2} \; ; \; f''\left(\dfrac{5\pi}{4}\right) = \sqrt{2}.$$

Since $f''\left(\dfrac{\pi}{4}\right) < 0$, $\left(\dfrac{\pi}{4}, \sqrt{2}\right)$ is a relative maximum. Since $f''\left(\dfrac{5\pi}{4}\right) > 0$, $\left(\dfrac{5\pi}{4}, -\sqrt{2}\right)$ is a relative minimum.

LIMITS AT INFINITY

Although horizontal and vertical asymptotes were previously discussed, we can use the notation of limits to continue the discussion. Horizontal asymptotes occur when the limit as x approaches $\pm\infty$ is a number. While questions of this type can be asked by asking about horizontal asymptotes, they can also be asked in the form of find $\lim\limits_{x \to \infty} f(x)$ or $\lim\limits_{x \to -\infty} f(x)$. In either case, although you can return to precalculus techniques to find the answer, you can also use the technique of dividing the numerator and denominator by the highest power of x found in the denominator. Note that $\lim\limits_{x \to +\infty} \dfrac{c}{x^n} = 0$ where c is a constant and n is a rational number greater than 0.

Example 56
Find each limit.

a. $\lim\limits_{x \to \infty} \dfrac{3x+1}{x^2 - 9}$ b. $\lim\limits_{x \to \infty} \dfrac{3x+1}{4x+5}$ c. $\lim\limits_{x \to -\infty} \dfrac{2x^2 + 5}{x - 7}$

Solution:

a. $\lim_{x\to\infty}\dfrac{3x+1}{x^2-9} = \lim_{x\to\infty}\dfrac{\dfrac{3x}{x^2}+\dfrac{1}{x}}{\dfrac{x^2}{x^2}-\dfrac{9}{x^2}} = \dfrac{0+0}{1-0} = 0$

b. $\lim_{x\to\infty}\dfrac{3x+1}{4x+5} = \lim_{x\to\infty}\dfrac{\dfrac{3x}{x}+\dfrac{1}{x}}{\dfrac{4x}{x}+\dfrac{5}{x}} = \dfrac{3+0}{4+0} = \dfrac{3}{4}$

c. $\lim_{x\to-\infty}\dfrac{2x^2+5}{x-7} = \lim_{x\to-\infty}\dfrac{\dfrac{2x^2}{x}+\dfrac{5}{x}}{\dfrac{x}{x}-\dfrac{7}{x}} = \lim_{x\to-\infty}\dfrac{2x+\dfrac{5}{x}}{1-\dfrac{7}{x}} = \dfrac{\left(\lim_{x\to-\infty}2x\right)+0}{1+0} = -\infty$,

which does not exist.

Therefore, this limit does not exist.

Infinite Limits

When the y values of a function increase or decrease without bound as the x values approach some number c, we can use infinite limits to describe this situation.

Example 57

Describe the behavior of the graph of $f(x) = \dfrac{4}{x-1}$ as x approaches 1.

Solution:

Since $x = 1$ is not in the domain of the function, there is no y value when $x = 1$. However, there is a vertical asymptote at $x = 1$. We can state:

$\lim_{x\to 1^+}\dfrac{4}{x-1} = +\infty$ and $\lim_{x\to 1^-}\dfrac{4}{x-1} = -\infty$. Examine the graph of $y = \dfrac{4}{x-1}$ and note the behavior around the vertical asymptote:

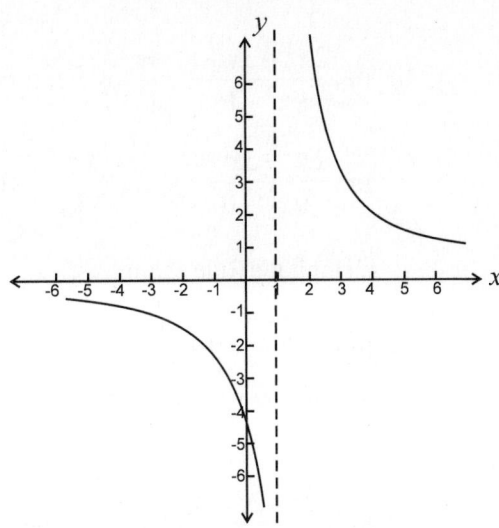

Or, you can reason in the following manner. When x approaches 1 from the right, the denominator is a small positive number. The fraction then becomes a large positive number, tending toward $+\infty$. When x approaches 1 from the left, the denominator is a small negative number. The fraction then becomes a large negative number, tending toward $-\infty$.

L'HÔPITAL'S RULE

L'Hôpital's Rule provides a way to find the limit of an indeterminate form of type $\frac{0}{0}$ or $\frac{\infty}{\infty}$. To find $\lim_{x \to a} \frac{f(x)}{g(x)}$, first replace x with a. If $\frac{0}{0}$ or $\frac{\infty}{\infty}$ is obtained, L'Hôpital's Rule can be applied, and $\lim_{x \to a} \frac{f'(x)}{g'(x)} = \lim_{x \to a} \frac{f(x)}{g(x)}$. Note here that a can be finite or infinite.

Example 58
Evaluate each limit.

a. $\lim_{x \to 3} \frac{x-3}{x^2-9}$ b. $\lim_{x \to \infty} \frac{2x^2-3x+1}{9x^2-16}$ c. $\lim_{x \to \infty} \frac{e^x}{x^3}$

Solutions:

a. Replacing x with 3 gives $\frac{0}{0}$, so L'Hôpital's Rule applies. Take the derivative of the numerator and the derivative of the denominator and reevaluate the limit: $\lim_{x \to 3} \frac{1}{2x} = \frac{1}{6}$. Note that if we used earlier techniques to find this limit, we would factor, cancel, and then substitute: $\lim_{x \to 3} \frac{x-3}{x^2-9} = \lim_{x \to 3} \frac{x-3}{(x-3)(x+3)} = \lim_{x \to 3} \frac{1}{x+3} = \frac{1}{6}$ so that the answers agree.

b. $\lim_{x \to \infty} \frac{2x^2 - 3x + 1}{9x^2 - 16} = \frac{\infty}{\infty}$ by direct substitution, so apply L'Hôpital's Rule:

$$\lim_{x \to \infty} \frac{2x^2 - 3x + 1}{9x^2 - 16} = \lim_{x \to \infty} \frac{4x - 3}{18x} \text{ which gives } \frac{\infty}{\infty}, \text{ so we apply}$$

L'Hôpital's Rule again:

$$\lim_{x \to \infty} \frac{4x - 3}{18x} = \lim_{x \to \infty} \frac{4}{18} = \frac{2}{9}.$$ If we use the rules discussed in the previous section, we would have:

$$\lim_{x \to \infty} \frac{2x^2 - 3x + 1}{9x^2 - 16} = \lim_{x \to \infty} \frac{\frac{2x^2}{x^2} - \frac{3x}{x^2} + \frac{1}{x^2}}{\frac{9x^2}{x^2} - \frac{16}{x^2}} = \frac{2 - 0 + 0}{9 - 0} = \frac{2}{9} \text{ (and of course,}$$

the answers agree.)

c. $\lim_{x \to \infty} \frac{e^x}{x^3} = \frac{\infty}{\infty}$, so we apply L'Hôpital's Rule:

$$\lim_{x \to \infty} \frac{e^x}{x^3} = \lim_{x \to \infty} \frac{e^x}{3x^2} = \lim_{x \to \infty} \frac{e^x}{6x} = \lim_{x \to \infty} \frac{e^x}{6} = \infty.$$ Note that L'Hôpital's Rule had to be applied three times before we arrived at the final answer.

VELOCITY, SPEED, AND ACCELERATION

Average velocity is the change in distance divided by the change in time, or $\frac{\Delta s}{\Delta t}$ where the function s gives the position of an object at a given time t. Instantaneous velocity, usually referred to as velocity, is the derivative of the position function, $v(t) = s'(t)$. Speed is the absolute value of velocity, $|v(t)|$ or $|s'(t)|$. Acceleration is the rate of change of velocity over time, or $a(t) = v'(t) = s''(t)$.

Example 59

A particle moves along a line so that at time t, where $0 \leq t \leq \frac{3\pi}{2}$, its position is given by $s(t) = 2\sin t - \frac{1}{2}t^2 + 8$.

a. What is the velocity of the particle when $t = \frac{\pi}{6}$?

b. What is the acceleration of the particle when $t = \frac{\pi}{6}$?

c. What is the velocity when the acceleration is 0?

d. Find the speed of the particle when $t = \frac{5\pi}{4}$.

DERIVATIVES

Solutions:

a. $v(t) = s'(t) = 2\cos t - t$ so $v\left(\dfrac{\pi}{6}\right) = 2\cos\dfrac{\pi}{6} - \dfrac{\pi}{6} \approx 1.21$

b. $a(t) = s''(t) = -2\sin t - 1$ so $a\left(\dfrac{\pi}{6}\right) = -2\sin\dfrac{\pi}{6} - 1 = -2$

c. The acceleration is 0 when $-2\sin t - 1 = 0$ or when $\sin t = -\dfrac{1}{2}$. Since $0 \le t \le \dfrac{3\pi}{2}$, this occurs only when $t = \dfrac{7\pi}{6}$. Now find velocity when

$t = \dfrac{7\pi}{6}$: $v\left(\dfrac{7\pi}{6}\right) = 2\cos\dfrac{7\pi}{6} - \dfrac{7\pi}{6}$. $= \sqrt{3} - \dfrac{7\pi}{6}$

d. Since speed equals the absolute value of velocity, find $\left|v\left(\dfrac{5\pi}{4}\right)\right|$. Now,

$v\left(\dfrac{5\pi}{4}\right) = 2\cos\dfrac{5\pi}{4} - \dfrac{5\pi}{4} = -5.34$, so speed equals 5.34.

VELOCITY AND ACCELERATION FOR VECTOR-VALUED FUNCTIONS

When the position of a function is defined by a position function $r(t) = f(t)\mathbf{i} + g(t)\mathbf{j}$, the derivative of $r(t)$, $r'(t)$, is called the velocity vector and the magnitude of velocity, $\|r'(t)\| = \sqrt{[x'(t)]^2 + [y'(t)]^2}$ gives the speed of the object at time t. $r''(t)$ gives the acceleration vector of the object at time t.

Example 60
Find the velocity vector, speed, and acceleration vector of a particle that moves along the plane curve c described by $r(t) = (t^2 + 2)\mathbf{i} + t\mathbf{j}$.

Solution:

$r'(t) = 2t\mathbf{i} + \mathbf{j}$ and $r'' = 2\mathbf{i}$. Note that when dealing with a vector-valued position function, the results of the first and second derivative are vectors.

Part III
INTEGRALS

RIEMANN SUM

Consider a curve $y = f(x)$ as shown in the figure below. We wish to find the area under the curve bounded by the x-axis and the ordinates $x = a$ and $x = b$ ($b > a$).

We can approximate the area by dividing the region into n vertical strips and adding up the area of each vertical strip to get the required area under the curve. For this, we divide $[a,b]$ into n subintervals. Let $x_0, x_1, x_2,..., x_n$ represent the end points of these intervals. For each interval $[x_{i-1}, x_i]$ choose a value c_i in the interval. Therefore the area under the curve can be approximated by the sum of areas of the rectangles drawn on top of each subinterval $[x_{i-1}, x_i]$ with height $f(c_i)$ as shown below.

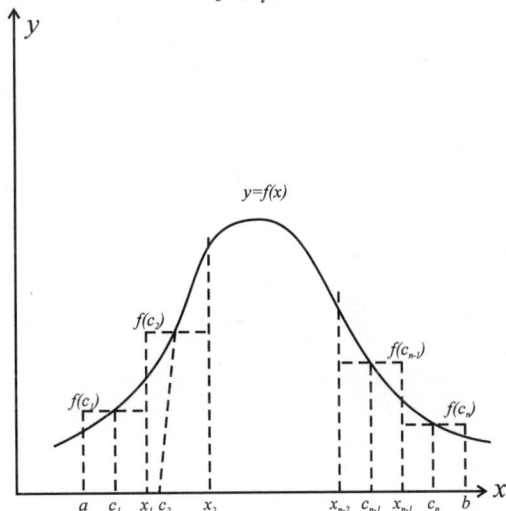

Let $\Delta x_i = x_i - x_{i-1}$
Then, the area under the curve can be approximated as:

$$A \approx f(c_1)*\Delta x_1 + f(c_2)*\Delta x_2 + ... + f(c_n)*\Delta x_n = \sum_{i=1}^{n} f(c_i)*\Delta x_i$$

The above sum is called a **Riemann Sum**.

To make this computation easier, we can choose the intervals to all have equal length, i.e. $\Delta x_i = \dfrac{b-a}{n}$, for all $i = 1, 2, 3, ..., n$.

Also, we can chose the value of c_i to be either the right-side (x_R), left-side (x_L) or the mid-point (x_M) of the interval $[x_{i-1}, x_i]$ as shown in case of a single strip in the following figure.

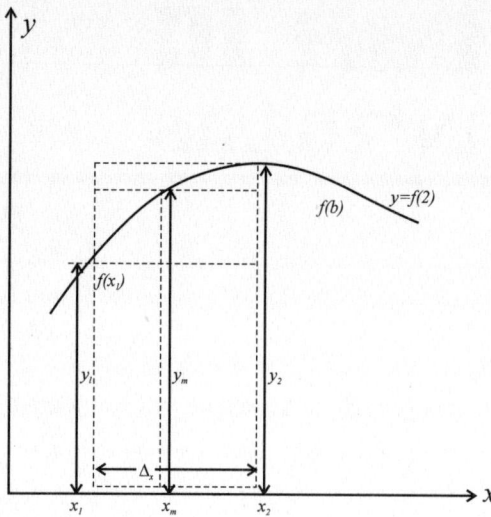

Example 1

Approximate the area bounded by the parabola $y = x^2$, the x-axis and the ordinates $x = 1$ and $x = 3$ using the left, right, and the midpoint evaluation points. Let the area be divided into ten rectangular strips of equal widths.

Solution:

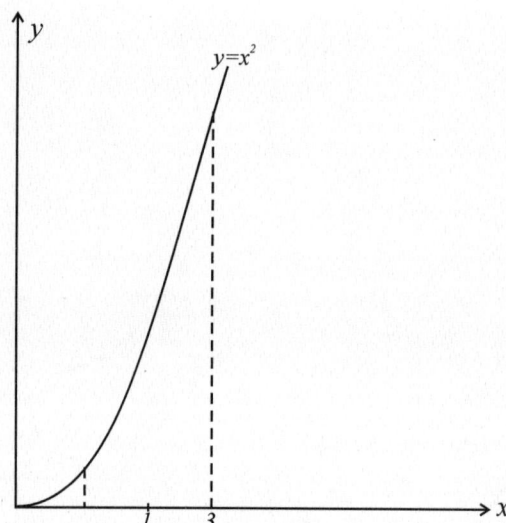

We have, $n = 10$, $a = 1$, and $b = 3$.

$$\Delta x = \frac{b-a}{n} = \frac{3-1}{10} = 0.2$$

a. Using left evaluation points:

For rectangle 1, $f(c_1) = f(1)$;

For rectangle 2, $f(c_2) = f(1.2)$; and so on.

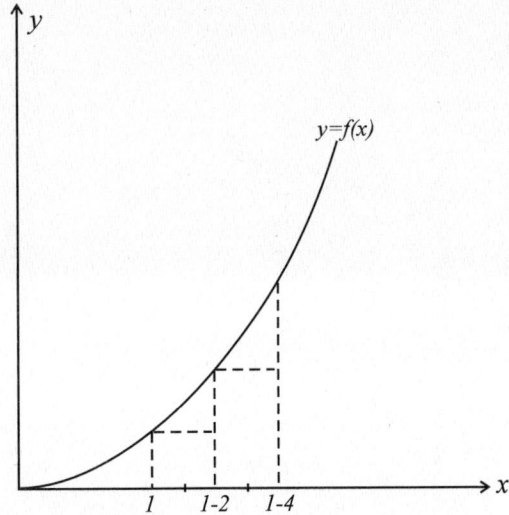

In general, for the ith rectangle, $f(c_i) = f(a + i * \Delta x)$, where $i = 0, 1, 2, \ldots, n-1$

$$\therefore A = \sum_{i=0}^{9} f(1 + i * 0.2) * 0.2$$

$= f(1) \cdot (0.2) + f(1.2) \cdot (0.2) + f(1.4) \cdot (0.2) + f(1.6) \cdot (0.2) + f(1.8) \cdot$

$(0.2) + f(2.0) \cdot (0.2) + f(2.2) \cdot (0.2) + f(2.4) \cdot (0.2) + f(2.6) \cdot$

$(0.2) + f(2.8) \cdot (0.2)$

$= 7.88$

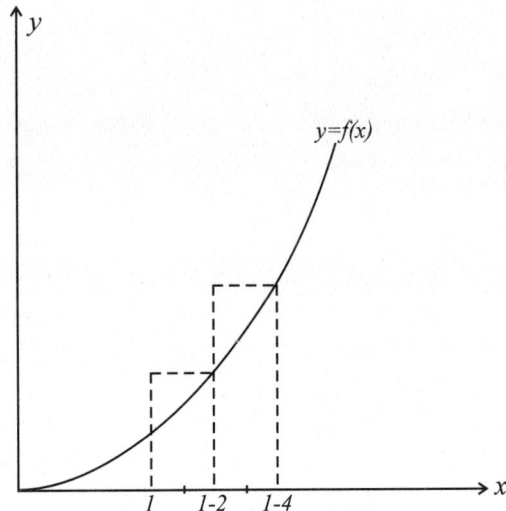

b. Using right evaluation points:

For rectangle 1, $f(c_1) = f(1.2)$;

For rectangle 2, $f(c_2) = f(1.4)$; and so on.

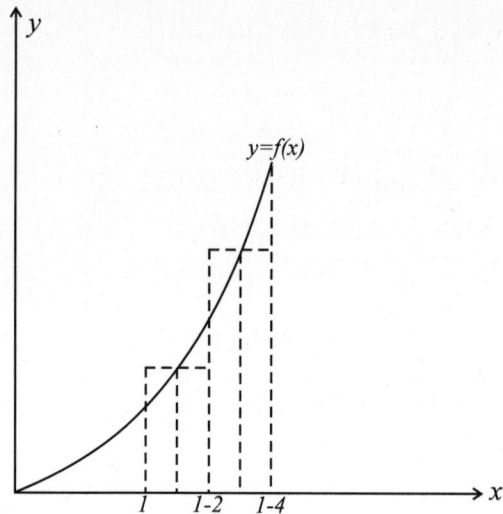

In general, for the ith rectangle, $f(c_i) = f(a + i * \Delta x)$, where $i = 1, 2, 3, ..., n-1, n$.

$$\therefore A = \sum_{i=1}^{10} f(1 + i * 0.2) * 0.2$$

$= f(1.2) \cdot (0.2) + f(1.4) \cdot (0.2) + f(1.6) \cdot (0.2) + f(1.8) \cdot (0.2) +$

$f(2.0) \cdot (0.2) + f(2.2) \cdot (0.2) + f(2.4) \cdot (0.2) + f(2.6) \cdot (0.2) +$

$f(2.8) \cdot (0.2) + f(3.0) \cdot (0.2)$

$= 9.48$

c. Using midpoint evaluation points:

For rectangle 1, $f(c_1) = f(1.1)$;

For rectangle 2, $f(c_2) = f(1.3)$; and so on.

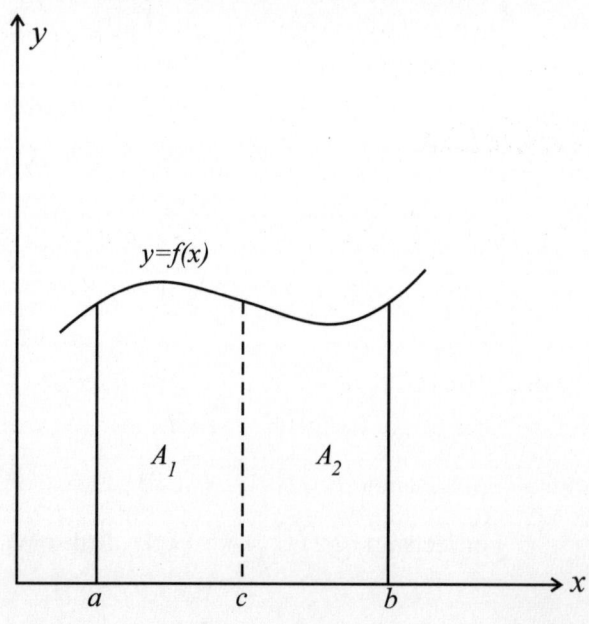

INTEGRALS

In general, for the ith rectangle,

$$f(c_i) = f(\frac{(a+(i-1)*\Delta x)+(a+i*\Delta x)}{2}), \text{ where } i = 1, 2, ..., n$$

$$\therefore A = \sum_{i=1}^{10} f(\frac{(1+(i-1)*0.2)+(1+i*0.2)}{2})*0.2$$

$$= f(1.1) \cdot (0.2) + f(1.3) \cdot (0.2) + f(1.5) \cdot (0.2) + f(1.7) \cdot (0.2) +$$

$$f(1.9) \cdot (0.2) + f(2.1) \cdot (0.2) + f(2.3) \cdot (0.2) + f(2.5) \cdot (0.2) +$$

$$f(2.7) \cdot (0.2) + f(2.9) \cdot (0.2)$$

$$= 8.66$$

Solve the following problems:

1. Approximate the area bounded by the line $y = x$, the x-axis and the ordinates $x = 5$ and $x = 8$, using the left, midpoint, and right evaluation points. Divide the area into 5 equal strips. Compare your results with those obtained geometrically (***Hint:*** The desired area represents a trapezium.) Which method of evaluation gives the most accurate results?

2. Approximate the area of the region bounded by the curve $y = 1/x$, the x-axis, and the ordinates $x = -1$ and $x = 3$ using the left evaluation points. Take $n = 10$.

3. Approximate the area bounded by the curve $x = 4 - y^2$, the y-axis, and the lines $y = 1$ and $y = 2$ using the midpoint evaluation points. Take $n = 6$. (***Hint***: The procedure is the same as in other cases, just that we have $x = f(y)$ in this case, and we change the formulae accordingly.)

THE DEFINITE INTEGRAL

Consider a curve $y = f(x)$. We have already seen that the area under the curve can be approximated by using a Riemann Sum given by $\sum_{i=1}^{n} f(c_i) * \Delta x$

Now, if we allow the width of each rectangle to be different, say $\Delta x_i = x_i - x_{i-1}$, for the ith rectangle, then we have:

The area, $A \approx \sum_{i=1}^{n} f(c_i) * \Delta x_i$, where c_i is any value in the interval $[x_{i-1}, x_i]$.

It is evident from the top figure on p. 132 that neither the right side nor the left side or the mid-point represent the true height of the rectangle. This problem can be taken care of by making Δx_i so small that the right side is almost equal to the left side of the rectangle. Thus, as Δx_i tends to zero, $f(x_L)$ approaches $f(x_R)$. In other words, we take the limiting value of the above expression as the maximum width tends to zero. Let's say it equals L.

Then, we have: $L = \lim_{\max \Delta x_i \to 0} \sum_{i=1}^{n} f(c_i) * \Delta x_i$

We now define a very important concept in calculus using the above expression.

Let $y = f(x)$ be defined on $[a,b]$ with $a < b$. Then the **Definite Integral** of f over the interval $[a, b]$ is given by $\int_{a}^{b} f(x)dx = \lim_{\max \Delta x_i \to 0} \sum_{i=1}^{n} f(c_i) * \Delta x_i$, wherever the limit exists.

The process of calculating an integral is called **Integration**. The numbers a and b are called the lower and upper bounds (or limits) of integration, respectively. If the limits exist, then the function is said to be integrable over the interval $[a,b]$. The function $f(x)$ is called the integrand.

Example 2
Calculate the value of $\int_{1}^{2} x\,dx$

Solution:

The above problem is basically same as that of obtaining the area under the curve $y = x$, bounded by the x-axis and the limits $x = 1$ and $x = 2$.

We first divide the interval $[1,2]$ into n equal subintervals, each of length

$\Delta x = \dfrac{2-1}{n} = \dfrac{1}{n}$

Taking left evaluation points, we have: $x_i = a + i * \Delta x$, $i = 0, 1, 2, \ldots, n-1$

Thus, $x_0 = a = 1$

$x_1 = 1 + \dfrac{1}{n}$

.

.

.

$x_{n-1} = 1 + \dfrac{n-1}{n}$

INTEGRALS

The areas of each of the n rectangles are:

$$f(x_0) * \Delta x = \frac{1}{n}$$

$$f(x_1) * \Delta x = (1 + \frac{1}{n})\frac{1}{n}$$

.

.

.

$$f(x_{n-1}) * \Delta x = (1 + \frac{n-1}{n})\frac{1}{n}$$

$\therefore$ The sum is:
$$\sum_{i=0}^{n-1} f(x_i) * \Delta x = \frac{1}{n} + \left(1 + \frac{1}{n}\right)\frac{1}{n} + \ldots + \left(1 + \frac{n-1}{n}\right)\frac{1}{n}$$

$$= \left[1 + \left(1 + \frac{1}{n}\right) + \ldots + \left(1 + \frac{n-1}{n}\right)\right]\frac{1}{n}$$

$$= 1 + \frac{1}{n^2}[0 + 1 + 2 + \ldots + (n-1)]$$

$$= 1 + \frac{1}{n^2}\left(\frac{n(n-1)}{2}\right)$$

$$= \frac{3}{2} - \frac{1}{2n}$$

$\therefore$ Area, $A = \lim_{\Delta x \to 0} \sum_{i=0}^{n-1} f(x_i) * \Delta x$

$$= \lim_{n \to \infty} \left(\frac{3}{2} - \frac{1}{2n}\right)$$

$$= \frac{3}{2}$$

Example 3

Calculate the value of $\int_0^1 x^2 dx$

Solution:

In this case, we have to calculate the area under the curve $y = x^2$ bounded by the x-axis, between $x = 0$ and $x = 1$.

We partition the interval $[0,1]$ into n equal subintervals, giving us the length of each subinterval, $\Delta x = \frac{1-0}{n} = \frac{1}{n}$

Taking the right evaluation points, we have:

$$x_i = a + i*\Delta x = 0 + i\left(\frac{1}{n}\right) = \frac{i}{n}$$

Therefore, for $i = 1, 2, 3, \ldots, n$ we have:

$$f(x_i) = f\left(\frac{i}{n}\right) = \left(\frac{i}{n}\right)^2$$

$$\sum_{i=1}^{n} f(x_i)*\Delta x$$

$$= \frac{1}{n^3} \sum_{i=1}^{n} i^2$$

$$= \frac{1}{n^3}\left(\frac{n(n+1)(2n+1)}{6}\right)$$

$$= \frac{(n+1)(2n+1)}{6n^2}$$

$$\therefore \int_0^1 x^2 dx = \lim_{\Delta x \to 0} \sum_{i=1}^{n} f(x_i)*\Delta x$$

$$= \lim_{n \to \infty} \frac{(n+1)(2n+1)}{6n^2}$$

$$= \lim_{n \to \infty} \frac{2n^2 + 3n + 1}{6n^2}$$

$$= \lim_{n \to \infty}\left(\frac{1}{3} + \frac{1}{2n} + \frac{1}{6n^2}\right)$$

$$= \frac{1}{3}$$

We saw how definite integrals can be used to express the area under a curve. We now show how definite integrals can be related to applications having nothing to do with areas.

Suppose that a particle is moving with a velocity $v = v(t) \geq 0$. If this velocity is constant, then the distance traveled between an initial time, t_0 and a final time, t_f may be given by $s = v*(t_f - t_0)$.

However, if the velocity over the period of time is changing, then the above formula does not work. In this case, we calculate the distance covered by a method similar to the one we used to calculate the area under a curve. The central idea is that even though the velocity is changing, it is almost constant over a very small period of time.

So, we divide the time interval $[t_0, t_f]$ into n equal subintervals and calculate the distance traveled by the particle in each of these subintervals. If n is taken to be very large, the velocity in each of these subintervals is almost constant and we end up with a very good approximation of the distance covered.

INTEGRALS

We have:

$$\Delta t = \frac{t_f - t_0}{n}.$$

If s_i is the distance covered in the ith time interval, then the approximate distance covered,

$$s_{app} = s_1 + s_2 + s_3 + \ldots + s_i + \ldots + s_n$$

$$= v(t_1)*\Delta t + v(t_2)*\Delta t + \ldots + v(t_i)*\Delta t + \ldots + v(t_n)*\Delta t$$

where t_i = a value in the ith time interval.

$$= \sum_{i=1}^{n} v(t_i)*\Delta t$$

Using our knowledge of definite integrals, we can obtain the actual distance covered as:

$$s = \lim_{\Delta t \to 0} \sum_{i=1}^{n} v(t_i)*\Delta t$$

$$= \int_{t_0}^{t_f} v(t)dt$$

Now, the total distance covered in the interval $[t_0, t_f]$ is also the distance covered at the end of time t_f minus the distance covered at the end of time t_0.

i.e. $s = s(t_f) - s(t_0)$

Also, from our knowledge of derivatives, we know that

$$v(t) = s'(t)$$

Therefore, we get:

$$s = s(t_f) - s(t_0) = \int_{t_0}^{t_f} s'(t)dt$$

In other words, the definite integral of rate of change of a quantity over an interval can be interpreted as the change of the quantity over the interval.

Basic Properties of Definite Integrals:

1. **Additive Property**: If two functions f and g are both integrable on $[a,b]$, then $f+g$ is integrable on $[a,b]$, and

$$\int_a^b (f+g) = \int_a^b f + \int_a^b g$$

Linearity Property: If a function f is integrable over $[a,b]$, and if k is any constant, then kf is integrable over $[a,b]$, and

$$\int_a^b kf = k\int_a^b f$$

2. We also have another important result of splitting up the intervals of definite integrals.

Consider the figure shown below. The area under the curve $y = f(x)$ may be given as

$$A = \int_a^b f(x)dx$$

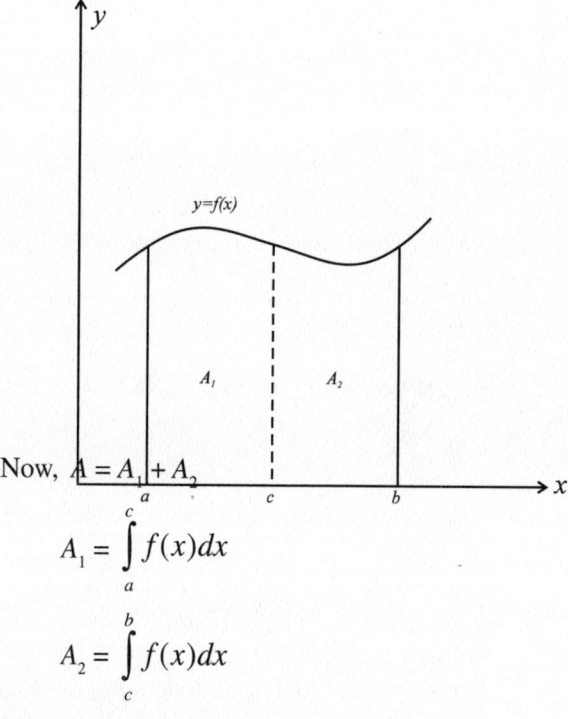

Now, $A = A_1 + A_2$

$$A_1 = \int_a^c f(x)dx$$

$$A_2 = \int_c^b f(x)dx$$

Therefore, we have:

$$A = \int_a^b f(x)dx = \int_a^c f(x)dx + \int_c^b f(x)dx$$

Note that $a \leq c \leq b$

INTEGRALS

Example 4
$$\int_1^3 (2x+3x^3)dx = \int_1^3 2x\,dx + \int_1^3 3x^3\,dx$$

Example 5
$$\int_a^b 4x^2\,dx = 4\int_a^b x^2\,dx$$

Example 6
$$\int_1^4 x^3\,dx = \int_1^3 x^3\,dx + \int_3^4 x^3\,dx$$

Obtain the values of the following definite integrals:

1. $\int_1^3 3x^2\,dx$

2. $\int_0^2 x^3\,dx$

3. $\int_1^4 (x^2+3x)dx$

APPLICATIONS OF INTEGRALS

We have already seen how we can use definite integrals to model problems on calculations of areas and distance covered along a straight line.
We now try to model some other practical examples using the concept of Riemann Sum and Definite Integrals.

Volume of a Solid:
The volume of any solid body with a constant cross-section of area A can be easily determined as $V = A*h$, where h is the height of the solid body. For example, the volume of a cylinder = $(\pi r^2)h = A*h$. This simple formula does not work if the cross-sectional area is changing along the height. In such cases, we modify our approach and use the incremental approach used in the previous topics.

Consider a solid shape of a variable cross-sectional area as shown on the next page.

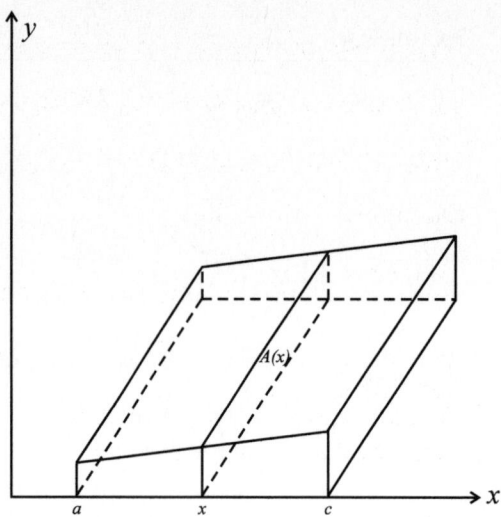

We divide the interval [a,b] into n subintervals with widths $\Delta x_1, \Delta x_2, \ldots, \Delta x_n$. We can imagine our solid to be sliced along these subintervals into smaller solid bodies of widths $\Delta x_1, \Delta x_2, \ldots, \Delta x_n$ respectively, as shown below.

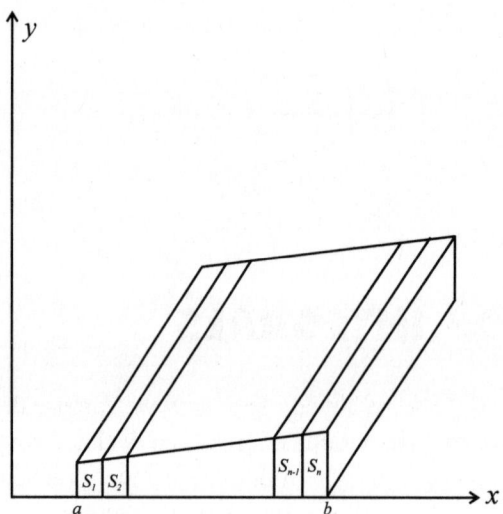

Consider a typical slice S_i. Let c_i be the value of the point at which we approximate the cross-sectional area of slice. Thus, the area may be approximated as,

$A_i = A(c_i)$, and the volume of the slice approximately becomes

$V_i = A(c_i) * \Delta x_i$

The net volume of the solid approximately is,

$$V_{app} = V_1 + V_2 + \ldots + V_i + \ldots V_n$$
$$= \sum_{i=1}^{n} A(c_i) * \Delta x_i$$

INTEGRALS

Now, if we increase the number of slices such that max $\Delta x_i \to 0$, the slices get thinner and thinner and we get a better and better approximation of the cross-sectional area and hence the volume.

Thus, $V = \displaystyle\lim_{\max \Delta x_i \to 0} \sum_{i=1}^{n} A(x_i) * \Delta x_i$

$= \displaystyle\int_a^b A(x)dx$, provided $A(x)$ is integrable.

Example 7

Find the volume of a solid obtained by revolving the curve $y = \sqrt{x}$ over the subinterval [1,3] about the x-axis.

Solution:

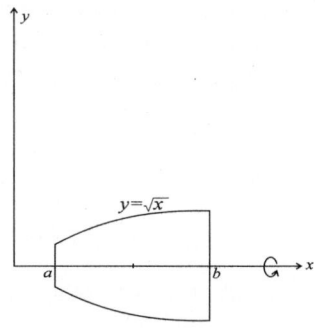

We have,

$V = \displaystyle\int_a^b A(x)dx$, where $a = 1$ and $b = 3$

To obtain $A(x)$, we take a cross-section across x-axis and find it's area.

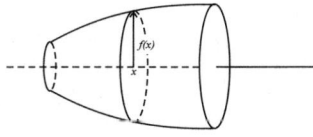

Notice that each cross-sectional piece is a circular disc with an area πr^2.

Therefore, the area of the circular disc at x would be $\pi(f(x))^2$, where $f(x)$ is the radius of the disc.

Therefore, $A(x) = \pi(f(x))^2$.

Thus, we have

$$V = \int_1^3 \pi(f(x))^2 \, dx$$

$$= \int_1^3 \pi(\sqrt{x})^2 \, dx$$

$$= \int_1^3 \pi x \, dx$$

$$= 4\pi$$

Solve the following problems:

1. Obtain the volume of a solid obtained by revolving the curve $x^2 + y^2 = 1$ over the interval $[-1,1]$ about the x-axis. What is the solid body obtained as a result of the revolution?

2. Calculate the volume of a truncated cone having the base and top radii of 3 units and 1 unit, respectively, and a length of 5 units along the x-axis.

3. Obtain the volume of a right pyramid having a square base of sides with length 1 unit. Take the height of the pyramid as 3 units. (**Hint:** Use the property of Similar Triangles to obtain the area of a square cross-section at any height y and integrate it over the entire height of the pyramid. Refer Fig. (b).)

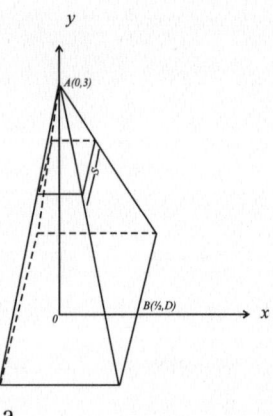
a.

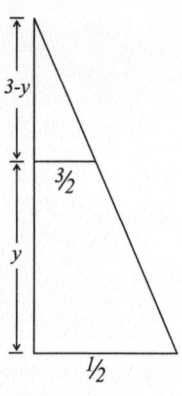
b.

INTEGRALS

Average Value of a Function:

Consider a curve $y = f(x)$ as shown below. The area under the curve over the interval $[a,b]$ is given as $A = \int_a^b f(x)dx$

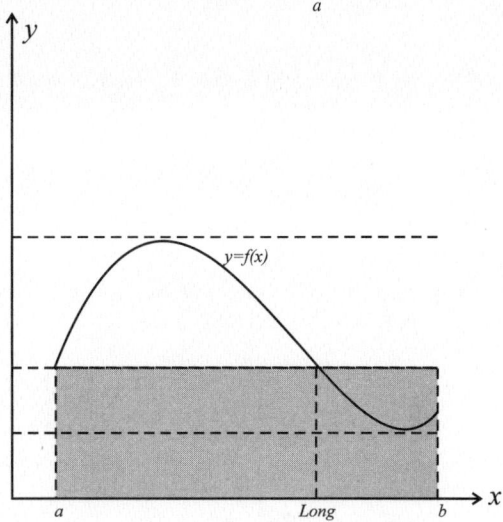

Also, let us assume a value $y_{avg} = f(x_{avg})$, where x_{avg} is in the interval $[a,b]$ such that the area under the curve $f(x)$ can be equated to a rectangle of width $(b - a)$ and height y_{avg}, as $A = (b - a) y_{avg}$.

Now, if y_{max} and y_{min} are the maximum and minimum values of $f(x)$ over $[a,b]$, then we can safely say that $y_{max} \geq y_{avg} \geq y_{min}$.

Therefore, $A = \int_a^b f(x)dx$

$= (b - a) y_{avg}$

This is called the *Mean Value Theorem of Integrals*, which basically states that if a function f is continuous in the closed interval $[a,b]$, then there is at least one number, x_{avg} in $[a,b]$ such that $\int_a^b f(x)dx = f(x_{avg})(b - a)$.

The number $f(x_{avg})$ is closely related to the concept of arithmetic average that we are so familiar with. To understand this, we first divide the interval $[a,b]$ into n equal subintervals of length $\Delta x = \dfrac{b-a}{n}$.

If we arbitrarily choose points $c_1, c_2, ..., c_n$ in successive subintervals, then the arithmetic average of $f(c_1), f(c_2), ..., f(c_n)$ would be

$$f(x_{avg}) = \frac{1}{n}\left[f(c_1) + f(c_2) + \ldots + f(c_n)\right]$$

$$= \frac{\Delta x}{b-a}\left[f(c_1) + f(c_2) + \ldots + f(c_n)\right]$$

$$= \frac{1}{b-a}\left[f(c_1)\Delta x + f(c_2)\Delta x + \ldots + f(c_n)\Delta x\right]$$

$$= \frac{1}{b-a}\sum_{i=1}^{n} f(c_i) * \Delta x$$

Taking limits as $\Delta x \to 0$, we have:

$$f_{avg} = \lim_{\Delta x \to 0} \frac{1}{b-a} \sum_{i=1}^{n} f(c_i) * \Delta x$$

$$= \frac{1}{b-a} \int_a^b f(x)dx$$

f_{avg} is known as the **average value** of the function $f(x)$ over the interval $[a,b]$.

Example 8
Find the average value of the function $f(x) = x^2$ over the interval $[0,1]$.

Solution:

$$f_{avg} = \frac{1}{b-a} \int_a^b f(x)dx$$

$$= \frac{1}{1-0} \int_0^1 x^2 dx$$

$$= 1\left(\frac{1}{3}\right) \quad \text{(See solved Example 3)}$$

$$= \frac{1}{3}$$

Solve the following problems:

1. Find the average value of the function $y = 2x$ over the interval $[5,8]$.

2. Obtain the height of a rectangle whose area is the same as that under the curve $y = 2x^3$ bounded by $x = 1$ and $x = 3$ and the x-axis. The width of the rectangle can be taken as the line segment between $x = 1$ and $x = 3$ (*Hint:* Equate the area of the rectangle to that of the region under the curve within the given limits and obtain the height of the rectangle.)

LENGTH OF A CURVE

In this section we will see how to find the arc length of a curve using definite integrals. We shall restrict our discussions to *smooth curves* only. A curve $y = f(x)$ is said to be smooth over an interval if it's derivative is continuous over the interval.

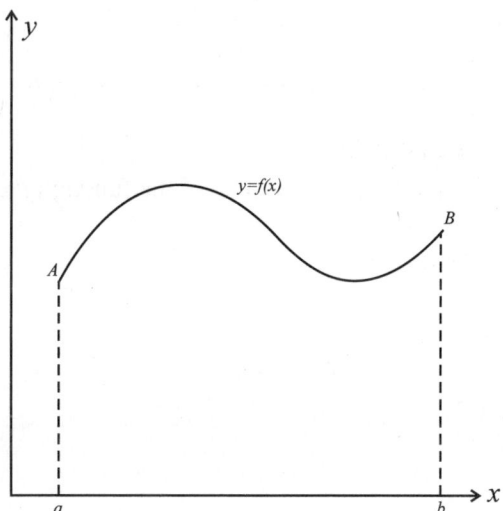

Consider a smooth curve given by the equation $y = f(x)$. We have to find the arc length, L of the curve over the interval $[a,b]$. The arc length of a curve can be visualized as the length of the string starting at point A and ending at point B.

We first divide the interval $[a,b]$ into n subintervals of length $\Delta x_1, \Delta x_2, \ldots, \Delta x_n$ by inserting points $x_1, x_2, x_3, \ldots, x_{n-1}$ between a and b. Let $Y_0, Y_1, Y_2, Y_3, \ldots, Y_{n-1}, Y_n$ be the points on the curve whose x-coordinates are $a, x_1, x_2, x_3, \ldots, x_{n-1}, b$ respectively, as shown in the top figure on the next page. By joining the points $Y_0, Y_1, Y_2, Y_3, \ldots, Y_{n-1}, Y_n$ on the curve by straight line segments, we obtain a polygonal path which we may approximate to the curve $y = f(x)$. The length of this polygonal path will approximate the length of the curve. And, as we increase the number of subintervals, the approximate length will approach the actual length of the curve.

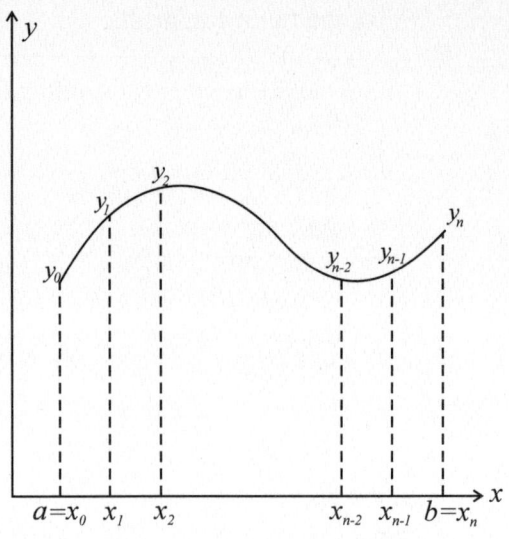

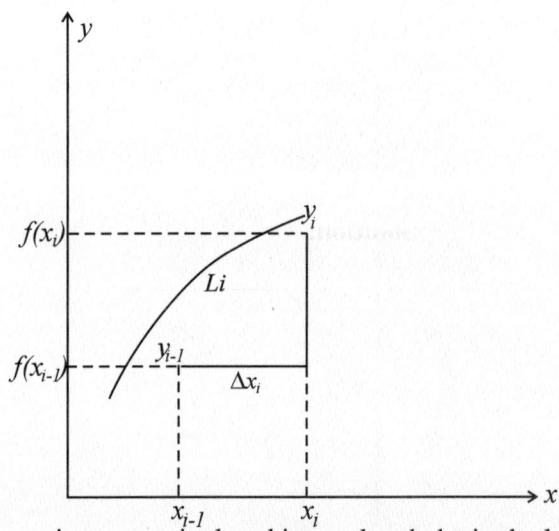

We zero in on one single subinterval and obtain the length of the arc in that subinterval. The length of the polygonal path using the *distance formula* becomes

$$L_i = \sqrt{(\Delta x_i)^2 + \left[f(x_i) - f(x_{i-1})\right]^2}$$

Now, by Mean Value Theorem of derivatives, we know that there is a point x_i^* between x_{i-1} and x_i such that $\dfrac{f(x_i) - f(x_{i-1})}{x_i - x_{i-1}} = f'(x_i^*)$

$$\therefore f(x_i) - f(x_{i-1}) = (x_i - x_{i-1}) f'(x_i^*)$$
$$= \Delta x_i \, f'(x_i^*)$$

INTEGRALS

$$\therefore L_i = \sqrt{(\Delta x_i)^2 + \left[\Delta x_i f'(x_i^*)\right]^2}$$

$$= \Delta x_i \sqrt{1 + \left[f'(x_i^*)\right]^2}$$

Over the entire interval, the length of the polygonal path becomes

$$L_{poly} = \sum_{i=1}^{n} \sqrt{1 + \left[f'(x_i^*)\right]^2} \, \Delta x_i$$

As we increase the number of subintervals such that max $\Delta x_i \to 0$, the length of the polygonal path approaches the length of the actual curve.

Therefore, $L = \lim_{\max \Delta x_i \to 0} \sum_{i=1}^{n} \sqrt{1 + \left[f'(x_i^*)\right]^2} \, \Delta x_i$

$$= \int_a^b \sqrt{1 + \left[f'(x)\right]^2} \, dx$$

Example 9

Find the length of the curve $y = x^{3/2}$ between $x = 0$ and $x = 1$.

Solution:

$$L = \int_a^b \sqrt{1 + \left[f'(x)\right]^2} \, dx$$

$$= \int_0^1 \sqrt{1 + \left(\frac{3}{2} x^{1/2}\right)^2} \, dx$$

$$= \int_0^1 \sqrt{1 + \frac{9x}{4}} \, dx$$

$$= \frac{1}{2} \int_0^1 (4 + 9x)^{1/2} \, dx$$

$$= \frac{1}{2} \left[\frac{1}{9} \cdot \frac{2}{3} \cdot (4 + 9x)^{3/2} \right]_0^1$$

$$= \frac{1}{27} \left[13^{3/2} - 4^{3/2} \right] = \frac{46.79}{27} = 1.73$$

Solve the following problems:

1. Obtain the arc length of the curve $y = 2x^2$ between $x = 1$ and $x = 4$.

2. Obtain the arc length of the curve $y = x^3$ between the points $x = -2$ and $x = 1$.

THE FUNDAMENTAL THEOREM OF CALCULUS AND THE ANTIDERIVATIVE

In the course of studying Definite Integrals, we developed an important result which stated that "the definite integral of rate of change of a quantity over an interval is equal to the change of that quantity over the interval."

i.e. $\int_a^b f'(x)dx = f(b) - f(a)$

where, $f'(x)$ is the derivative of $f(x)$. The above relation between $f(x)$ and $f'(x)$ can also be stated by saying that $f(x)$ is an antiderivative of $f'(x)$.

If F is a function defined on $[a,b]$ such that $F'(x) = f(x)$ for all x in $[a,b]$ then F is called an **antiderivative** or **indefinite integral** of f. Also, we can define the indefinite integral of f, denoted as $\int f(x)dx$, as the set of all possible antiderivatives of f.

Note that in the above definition we see that since F is not unique, it is defined as "*an*" antiderivative of f. This can easily be seen through the following example.

If $f(x) = 6x$, then both $F(x) = 3x^2 + 5$ and $G(x) = 3x^2 + 8$ are antiderivatives of $f(x)$ since $F'(x) = 6x = G'(x)$.

In general, we can say that for any function $F(x) = 3x^2 + C$, the derivative is $f(x) = 6x$; or the antiderivative of $f(x)$, in general, is of the form $F(x) + C$.

Thus, $\int f(x)dx = F(x) + C$ where F is any antiderivative of f and C is an arbitrary constant called the **Constant of Integration**.

The Fundamental Theorem of Calculus:

If f is continuous on $[a,b]$ and if F is an antiderivative of f on $[a,b]$, then

INTEGRALS

$$\int_a^b f(x)dx = F(b) - F(a)$$

Now,

$$\int_a^b f(x)dx = F(b) - F(a)$$

$$= [F(b) + C] - [F(a) + C]$$

$$= \left[\int f(x)dx\right]_a^b$$

Thus, we see that to evaluate any definite integral, all we have to do is evaluate any antiderivative at the end points of the interval and subtract one from the other accordingly.

Through the Fundamental Theorem of Calculus, we have established a *connection* between the derivative, antiderivative, and the definite integral.

TECHNIQUES OF ANTIDIFFERENTIATION

By definition of antiderivatives, we know that if $f(x) = F'(x)$, then

$$\int f(x)dx = F(x) + C$$

Consider the function $f(x) = x^n$

$$\frac{d}{dx}x^n = nx^{n-1}$$

$$\int nx^{n-1}dx = x^n + C$$

Or, $\int x^n dx = \dfrac{x^{n+1}}{n+1} + C$

Following are the antiderivatives of some basic functions.

$\dfrac{d(x)}{dx} = 1$	$\int dx = x + C$
$\dfrac{d}{dx}(\dfrac{x^{n+1}}{n+1}) = x^n$	$\int x^n dx = \dfrac{x^{n+1}}{n+1} + C$
$\dfrac{d(\sin x)}{dx} = \cos x$	$\int \cos x\, dx = \sin x + C$
$\dfrac{d(-\cos x)}{dx} = \sin x$	$\int \sin x\, dx = -\cos x + C$
$\dfrac{d(\tan x)}{dx} = \sec^2 x$	$\int \sec^2 x\, dx = \tan x + C$
$\dfrac{d(-\cot x)}{dx} = \csc^2 x$	$\int \csc^2 x\, dx = -\cot x + C$
$\dfrac{d(\sec x)}{dx} = \sec x \tan x$	$\int \sec x \tan x\, dx = \sec x + C$
$\dfrac{d(-\csc x)}{dx} = \csc x \cot x$	$\int \csc x \cot x\, dx = -\csc x + C$
$\dfrac{d(\sin^{-1} x)}{dx} = \dfrac{1}{\sqrt{1-x^2}}$	$\int \dfrac{1}{\sqrt{1-x^2}}\, dx = \sin^{-1} x + C$
$\dfrac{d(\cos^{-1} x)}{dx} = \dfrac{-1}{\sqrt{1-x^2}}$	$\int \dfrac{-1}{\sqrt{1-x^2}}\, dx = \cos^{-1} x + C$
$\dfrac{d(\tan^{-1} x)}{dx} = \dfrac{1}{1+x^2}$	$\int \dfrac{1}{1+x^2}\, dx = \tan^{-1} x + C$
$\dfrac{d(\ln|x|)}{dx} = \dfrac{1}{x}$	$\int \dfrac{1}{x}\, dx = \ln|x| + C$
$\dfrac{d(e^x)}{dx} = e^x$	$\int e^x\, dx = e^x + C$
$\dfrac{d(a^x)}{dx} = a^x \ln a$	$\int a^x\, dx = \dfrac{a^x}{\ln a} + C$

Example 10

Evaluate $\int 4x^3 dx$

Solution:

$$\int 4x^3 dx = 4\int x^3 dx$$
$$= 4\frac{x^4}{4} + C$$
$$= x^4 + C$$

Example 11

Evaluate $\int (x + x^2) dx$

Solution:

$$\int (x + x^2) dx = \int x\,dx + \int x^2 dx$$
$$= \frac{x^2}{2} + \frac{x^3}{3} + C$$

Example 12

Evaluate $\int \frac{\sin x}{\cos^2 x} dx$

Solution:

$$\int \frac{\sin x}{\cos^2 x} dx = \int \tan x \sec x\, dx \quad \text{(using trigonometric identities to first simplify)}$$
$$= \sec x + C$$

Example 13

Evaluate $\int \frac{x^4 + 2x^2 + 1}{x^2} dx$

Solution:

$$\int \frac{x^4 + 2x^2 + 1}{x^2} dx = \int \left(\frac{x^4}{x^2} + \frac{2x^2}{x^2} + \frac{1}{x^2}\right) dx$$
$$= \int \left(x^2 + 2 + \frac{1}{x^2}\right) dx$$
$$= \frac{x^3}{3} + 2x - \frac{1}{x} + C$$

Example 14
Obtain the area under the curve $y = \sin x$, bounded by the x-axis and $x = 0$ and $\frac{\pi}{2}$.

Solution:
The required area can be obtained by evaluating the following:

$$A = \int_0^{\pi/2} \sin x\, dx$$

$$= -\cos x \Big]_0^{\pi/2}$$

$$= -\cos \pi/2 + \cos 0$$

$$= 1$$

Solve the following:

1. $\int (x+1)(x+2)\,dx$

2. $\int (3x^2 + 6x)\,dx$

3. $\int \frac{x^5 + 6x^4 + x^2}{x^4}\,dx$

4. $\int \frac{x^2 + 7x + 6}{(x+1)}\,dx$

5. $\int \frac{\cos x}{\sin^2 x}\,dx$

6. $\int \frac{1}{\cos^2 x}\,dx$

7. Find the area under the curve $y = 3x^2 + 2x + 1$ over the interval [1,4] bounded by the x-axis.

8. Obtain the area bounded by the curve $y = 2\cos x$, the x-axis and the interval $[0, \pi/2]$.

9. Obtain the area bounded by the curve $y = 3\sin x$, the x-axis and the interval $[0, \pi]$.

INTEGRATION BY SUBSTITUTION

In this section, we shall discuss a technique of integration, which can often be used to transform a complicated integration problem into a simpler one. It is useful in cases where we cannot use the formulae of integration directly and have to modify our problem to fit one of the standard formulae.

For example, we know the integral of x^3 or that of $\sin x$, but what do we do for obtaining the integral of $(3x + 1)^3$ or $\sin(2x - 3)$?

In such cases we use a technique called **Integration By Substitution**. To show exactly how we do this, we start with a familiar example of the chain rule of derivatives.

Consider the function $f(x) = \dfrac{[g(x)]^{n+1}}{n+1}$

Then, $\dfrac{d}{dx}f(x) = \dfrac{d}{dx}\left(\dfrac{[g(x)]^{n+1}}{n+1}\right)$

$= [g(x)]^n g'(x)$

$\therefore \int [g(x)]^n g'(x) dx = \int \dfrac{d}{dx}\left(\dfrac{[g(x)]^{n+1}}{n+1}\right)$

$= \dfrac{[g(x)]^{n+1}}{n+1} + C$

If we let $u = g(x)$, then $du = g'(x)dx$.

Therefore, we have:

$$\int u^n du = \dfrac{u^{n+1}}{n+1} + C$$

Thus, we observe that if we can modify our integral from $\int [g(x)]^n g'(x) dx$ to $\int u^n du$, we are able to evaluate it easily.

In general, if we have to evaluate $\int [g(x)]g'(x)dx$, we substitute as follows:

$u = g(x) \Rightarrow du = g'(x)dx$, which finally gives us $\int f(u)du$.

Note that the final integral is in terms of u only.

With a *good choice* of u, we can obtain a very simple final integral to be evaluated. And, finally, after we evaluate the new integral, we replace the u by $g(x)$.

Example 15

Evaluate $\int 2x(x^2+1)^3 \, dx$

Solution:

Here, if we let $u = x^2 + 1$, we get $du = 2x\,dx$

$$\therefore \int 2x(x^2+1)^3 \, dx = \int (x^2+1)^3 \, 2x\,dx$$

$$= \int u^3 \, du$$

$$= \frac{u^4}{4} + C$$

$$= \frac{(x^2+1)^4}{4} + C$$

Example 16

Evaluate $\int \frac{(\sqrt{x}+3)^5}{\sqrt{x}} \, dx$

Solution:

Here we put $u = \sqrt{x} + 3$

$$\therefore du = \frac{1}{2\sqrt{x}} \, dx$$

or, $2du = \frac{1}{\sqrt{x}} \, dx$

$$\therefore \int \frac{(\sqrt{x}+3)^5}{\sqrt{x}} \, dx = \int (u^5) 2 \, du$$

$$= \int 2u^5 \, du$$

$$= \frac{2u^6}{6} + C$$

$$= \frac{(\sqrt{x}+3)^6}{3} + C$$

INTEGRALS

Example 17

Evaluate $\int \dfrac{2x(x^2+3)}{(x^2+1)^4}dx$

Solution:

Note a small difference in the technique in this example.

Here, we choose $u = (x^2 + 1)$
∴ $du = 2xdx$.

Also, $x^2 + 3 = u + 2$

$$\therefore \int \dfrac{2x(x^2+3)}{(x^2+1)^4}dx = \int \dfrac{(u+2)}{u^4}du$$

$$= \int \left(\dfrac{1}{u^3} + \dfrac{2}{u^4}\right)du$$

$$= \dfrac{-1}{2u^2} - \dfrac{2}{3u^3} + C$$

$$= \dfrac{-1}{2(x^2+1)^2} - \dfrac{2}{3(x^2+1)^3} + C$$

Example 18

Evaluate $\int \cos(3x+1)dx$

Solution:

Let $u = 3x + 1 \Rightarrow du = 3dx$ or $du/3 = dx$

$$\therefore \int \cos(3x+1)dx = \int \dfrac{\cos u}{3}du$$

$$= \dfrac{\sin u}{3} + C$$

$$= \dfrac{\sin(3x+1)}{3} + C$$

Note: In general, if $\int f(x)dx = F(x) + C$, then

$$\int f(ax+b)dx = \dfrac{F(ax+b)}{a} + C$$

Evaluate the following:

1. $\int \cos 5x \, du$

2. $\int \dfrac{1}{(3x-5)^6} \, dx$

3. $\int x^4 \sqrt{2-5x^5} \, dx$

4. $\int x^2 \sqrt{x-1} \, dx$

5. $\int (x + \sec^2 3x) \, dx$

6. $\int \sin^6 x \cos x \, dx$

7. $\int \sec^3 x \tan x \, dx$

8. $\int \sin(\sin \theta) \cos \theta \, d\theta$

Integration By Parts

In this section, we shall study a very powerful technique of integration called **integration by parts**. It has been derived from the product rule of differentiation.

If u and v are differentiable functions, then by product rule, we have

$$\frac{d}{dx}(uv) = u\frac{dv}{dx} + v\frac{du}{dx}.$$

Integrating both sides, we get

$$uv = \int u \, dv + \int v \, du$$

Rearranging the terms, we get

$$\int u \, dv = uv - \int v \, du$$

This is the formula for **Integration By Parts**.

Example 19

Evaluate $\int x \cos x \, dx$

INTEGRALS

Solution:

Here, we set $u = x$, so that $du = dx$ and the term in x vanishes.

Setting $\cos x\, dx = dv \Rightarrow v = \sin x$

$$\therefore \int x \cos x\, dx = uv - \int v\, du$$

$$= x \sin x - \int \sin x\, dx$$

$$= x \sin x + \cos x + C$$

Example 20

Evaluate $\int x^2 e^x\, dx$

Solution:

Here, let $u = x^2 \Rightarrow du = 2x\, dx$

$$dv = e^x dx \Rightarrow v = e^x$$

$$\therefore \int x^2 e^x\, dx = uv - \int v\, du$$

$$= x^2 e^x - \int e^x (2x\, dx)$$

$$= x^2 e^x - 2 \int x e^x\, dx$$

Now, for $2 \int x e^x\, dx$, we can integrate by parts a second time.

Thus, we take $u = x \Rightarrow du = dx$

$$dv = e^x dx \Rightarrow v = e^x$$

$$\therefore 2 \int x e^x\, dx = 2\left(x e^x - \int e^x dx \right)$$

$$= 2(x e^x - e^x)$$

$$\therefore \int x^2 e_x\, dx = x^2 e^x - 2x e^x + 2e^x + C$$

Example 21

Evaluate $\int \tan^{-1} x\, dx$

Solution:

Here, let $u = \tan^{-1} x \Rightarrow du = \dfrac{1}{1+x^2} dx$

$$dv = dx \Rightarrow v = x$$

$$\therefore \int \tan^{-1} x\, dx = x \tan^{-1} x - \int \dfrac{x}{1+x^2} dx$$

Now, for $\int \dfrac{x}{1+x^2} dx$, we use u-substitution, thus taking

$u = 1 + x^2 \Rightarrow du = 2xdu$

$$\therefore \int \dfrac{x}{1+x^2} dx = \int \dfrac{1}{2u} du$$

$$= \dfrac{1}{2} \ln|u| + C$$

$$= \dfrac{1}{2} \ln|1+x^2| + C$$

$$\therefore \int \tan^{-1} x dx = x \tan^{-1} x - \dfrac{1}{2} \ln|1+x^2| + C$$

Example 22

Evaluate $\int \cos^n x dx$

Solution:

We take $\quad u = \cos^{n-1} x \Rightarrow du = (n-1)\cos^{n-2} x(-\sin x) dx$

$dv = \cos x dx \Rightarrow v = \sin x$

$$\therefore \int \cos^n x dx = (\sin x)(\cos^{n-1} x) - \int (-\sin^2 x)(n-1)\cos^{n-2} x dx$$

$$= (\sin x)(\cos^{n-1} x) + (n-1)\int \sin^2 x \cos^{n-2} x dx$$

$$= (\sin x)(\cos^{n-1} x) + (n-1)\int (1-\cos^2 x)\cos^{n-2} x dx$$

$$= (\sin x)(\cos^{n-1} x) + (n-1)\int (\cos^{n-2} x - \cos^n x) dx$$

$$= (\sin x)(\cos^{n-1} x) + (n-1)\int \cos^{n-2} x dx - (n-1)\int \cos^n x dx$$

$$\therefore n\int \cos^n x dx = (\sin x)(\cos^{n-1} x) + (n-1)\int \cos^{n-2} x dx$$

$$\therefore \int \cos^n x dx = \dfrac{1}{n}(\sin x)(\cos^{n-1} x) + \dfrac{n-1}{n}\int \cos^{n-2} x dx$$

The above formula, called the **Reduction Formula**, reduces the exponent of $\cos^n x$ by 2. On repeatedly applying it, we can eventually express

$\int \cos^n x dx$ in terms of

$\int \cos x dx = \sin x + C$, if n is odd, and

$\int \cos^0 x \, dx = \int dx = x + C$, if n is even.

INTEGRALS

Similarly, we also have

$$\int \sin^n x\, dx = \frac{-1}{n}(\cos x)(\sin^{n-1} x) + \frac{n-1}{n}\int \sin^{n-2} x\, dx$$

Example 23

Evaluate $\int \cos^3 x\, dx$

Solution:

For $n = 3$, we have:

$$\int \cos^3 x\, dx = \frac{1}{3}(\sin x)(\cos^2 x) + \frac{2}{3}\int \cos x\, dx$$

$$= \frac{1}{3}(\sin x)(\cos^2 x) + \frac{2}{3}\sin x + C$$

Solve the following problems:

1. $\int xe^{-x}\, dx$
2. $\int \ln x\, dx$
3. $\int \sin^{-1} x\, dx$
4. $\int \cos^4 x\, dx$
5. $\int \sin^3 x\, dx$
6. $\int x^3 e^{x^2}\, dx$ (*Hint:* Take $u = x^2$ and $dv = xe^{x^2}$)

* Integration of Rational Functions Using Partial Fractions:

A rational function is the quotient of two polynomials. If it is of the form $\frac{P(x)}{Q(x)}$ and the degree of $P(x)$ is less than that of $Q(x)$, then it is called a **proper** rational function, or else it is called an **improper** rational function.

It can be proven that any proper rational function can be expressed as a sum of terms called **partial fractions** having the form: $\frac{A}{(ax+b)^n}$ or $\frac{Bx+C}{(ax^2+bx+c)^n}$, where $(ax^2 + bx + c)$ is irreducible.

For example, $\dfrac{4}{x^2-1} = \dfrac{2}{x-1} - \dfrac{2}{x+1}$

To obtain the partial fractions of a rational function, we follow the following steps:

(i) Factorize $Q(x)$ into linear and irreducible quadratic factors.

(ii) For each factor of the form $(ax + b)^j$, we introduce j terms as follows:

$$\dfrac{A_1}{ax+b}, \dfrac{A_2}{(ax+b)^2}, \ldots, \dfrac{A_j}{(ax+b)^j}, \text{ where } A_1, A_2, \ldots, A_j \text{ are constants to be determined.}$$

(iii) For each factor of the form $(ax^2 + bx + c)^k$, we introduce k terms as follows:

$$\dfrac{A_1 x + B_1}{(ax^2+bx+c)}, \dfrac{A_2 x + B_2}{(ax^2+bx+c)^2}, \ldots, \dfrac{A_k x + B_k}{(ax^2+bx+c)^k}, \text{ where}$$

$A_1, A_2, \ldots, A_k, B_1, B_2, \ldots, B_k$ are constants to be determined.

Once we have the partial fractions, we can evaluate the resultant integral by standard techniques.

Example 24

Evaluate $\displaystyle\int \dfrac{1}{x^2+x-2}\,dx$

Solution:

We have:

$$\dfrac{1}{x^2+x-2} = \dfrac{1}{(x-1)(x+2)}$$

As per the rule, the factor $(x - 1)$ introduces a term $\dfrac{A}{(x-1)}$ and the factor $(x + 2)$ introduces a term $\dfrac{B}{(x+2)}$

$$\therefore \dfrac{1}{x^2+x-2} = \dfrac{1}{(x-1)(x+2)} = \dfrac{A}{x-1} + \dfrac{B}{x+2}$$

INTEGRALS

To determine A and B, we multiply both sides by $(x-1)(x+2)$ to get:

$$1 = A(x+2) + B(x-1)$$
$$= Ax + 2A + Bx - B$$
$$= (A+B)x + (2A-B)$$

Equating coefficients, we have:
  $A + B = 0$, and
  $2A - B = 1$

Solving simultaneously, we get: $A = \dfrac{1}{3}$, and $B = \dfrac{-1}{3}$

$$\therefore \frac{1}{(x^2+x-2)} = \frac{1}{3(x-1)} - \frac{1}{3(x+2)}$$

$$\therefore \int \frac{1}{x^2+x-2}\,dx = \int \frac{1}{3(x-1)}\,dx - \int \frac{1}{3(x+2)}\,dx$$

$$= \frac{1}{3}\ln|x-1| - \frac{1}{3}\ln|x+2| + C$$

$$= \frac{1}{3}\ln\left|\frac{x-1}{x+2}\right| + C$$

Example 25

Evaluate $\int \dfrac{2x+4}{x^3-2x^2}\,dx$

Solution:

We have:

$$\frac{2x+4}{x^3-2x^2} = \frac{2x+4}{x^2(x-2)}$$

We introduce two terms for x^2 as follows: $\dfrac{A}{x} + \dfrac{B}{x^2}$, and

one term for $(x-2)$ as follows: $\dfrac{C}{x-2}$

Thus, we have the following which we solve for A, B and C:

$$\frac{2x+4}{x^3-2x^2} = \frac{A}{x} + \frac{B}{x^2} + \frac{C}{x-2}$$

Multiplying by $(x^3 - 2x^2)$, we get:
$$2x + 4 = Ax(x - 2) + B(x - 2) + Cx^2$$
$$= (C + A)x^2 + (B - 2A)x - 2B$$

Equating corresponding coefficients, we have:
$C + A = 0$
$B - 2A = 2$
$-2B = 4$

Solving the above system of equations simultaneously, we have:
$A = -2,$
$B = -2,$ and
$C = 2$

$$\therefore \frac{2x + 4}{x^3 - 2x^2} = \frac{-2}{x} - \frac{2}{x^2} + \frac{2}{x - 2}$$

$$\therefore \int \frac{2x + 4}{x^3 - 2x^2} dx = -2\int \frac{1}{x} dx - 2\int \frac{1}{x^2} dx + 2\int \frac{1}{x - 2} dx$$

$$= -2\ln|x| + \frac{2}{x} + 2\ln|x - 2| + C$$

$$= 2\ln\left|\frac{x - 2}{x}\right| + \frac{2}{x} + C$$

Example 26

Evaluate $\int \frac{3x^4 + 3x^3 - 5x^2 + x - 1}{x^2 + x - 2} dx$

Solution:

Here, we notice that the above rational function is improper. Hence, we cannot obtain its partial fractions.
In such a case, we perform a long division to obtain:

$$\frac{3x^4 + 3x^3 - 5x^2 + x - 1}{x^2 + x - 2} = (3x^2 + 1) + \frac{1}{x^2 + x - 2}$$

$$\therefore \int \frac{3x^4 + 3x^3 - 5x^2 + x - 1}{x^2 + x - 2} dx = \int (3x^2 + 1)dx + \int \frac{1}{x^2 + x - 2} dx$$

$$= (x^3 + x) + \frac{1}{3}\ln\left|\frac{x - 1}{x + 2}\right| + C$$

(refer to Example 24)

Solve the following:

1. $\int \dfrac{1}{x^2+8x+7}\,dx$

2. $\int \dfrac{1}{x(x-1)^2}\,dx$

3. $\int \dfrac{1}{x^3-x}\,dx$

4. $\int \dfrac{x+2}{(x-1)(x-2)(x-3)}\,dx$

5. $\int \dfrac{x^2-4}{x-1}\,dx$

6. $\int \dfrac{x^3}{x^2-x-6}\,dx$

IMPROPER INTEGRALS

In our definition of **Definite Integral**, $\int_a^b f(x)\,dx$, we made two assumptions:

(i) The interval [a,b] is bounded, and
(ii) f is bounded on the interval [a,b].

Integrals that have unbounded intervals of integration or in which the integrand is unbounded within the interval are called **Improper Integrals**.
In this section, we shall see how to evaluate improper integrals.

Integrals with unbounded intervals of integration:

If f is continuous on the interval $[a, \infty)$, then we can define

$$\int_a^\infty f(x)\,dx = \lim_{k\to\infty} \int_a^k f(x)\,dx$$

If f is continuous on the interval $(-\infty, b]$, then we can define

$$\int_{-\infty}^b f(x)\,dx = \lim_{k\to-\infty} \int_k^b f(x)\,dx$$

If the limit exists, then the improper integral is said to **converge** and the value of the limit is assigned to the integral. If the limit does not exist, then the improper integral is said to **diverge** and the integral is not assigned a value.

Example 27

Evaluate $\int_1^\infty \frac{1}{x^2} dx$

Solution:

$$\int_1^\infty \frac{1}{x^2} dx = \lim_{k \to \infty} \int_1^k \frac{1}{x^2} dx$$

$$= \lim_{k \to \infty} \left[\frac{-1}{x} \right]_1^k$$

$$= \lim_{k \to \infty} \left(1 - \frac{1}{k} \right)$$

$$= 1$$

Example 28

Evaluate $\int_0^\infty \cos x\, dx$

Solution:

$$\int_0^\infty \cos x\, dx = \lim_{k \to \infty} \int_0^k \cos x\, dx$$

$$= \lim_{k \to \infty} [\sin x]_0^k$$

Now, $-1 \leq \sin k \leq 1$ for all values of k. Thus we see that the above limit does not converge at all. Hence the integral diverges.

Example 29

Evaluate $\int_{-\infty}^0 e^x dx$

Solution:

$$\int_{-\infty}^{0} e^x \, dx = \lim_{k \to -\infty} \int_{k}^{0} e^x \, dx$$

$$= \lim_{k \to -\infty} \left[e^x \right]_{k}^{0}$$

$$= \lim_{k \to -\infty} (1 - e^k)$$

$$= 1$$

Example 30

Evaluate $\int_{0}^{\infty} \dfrac{1}{x^{1/2}} \, dx$

Solution:

$$\int_{0}^{\infty} \frac{1}{x^{1/2}} \, dx = \lim_{k \to \infty} \int_{0}^{k} \frac{1}{x^{1/2}} \, dx$$

$$= \lim_{k \to \infty} \left[\frac{-2x^{1/2}}{2} \right]_{0}^{k}$$

$$= \lim_{k \to \infty} \left[\frac{1}{2} - \frac{2k^{1/2}}{2} \right]$$

Here, the above limit approaches ∞ as $k \to \infty$. Thus, the integral diverges.

Note: In general, $\int_{-\infty}^{\infty} f(x)dx = \int_{-\infty}^{c} f(x)dx + \int_{c}^{\infty} f(x)dx$, where c lies in the interval $(-\infty, \infty)$

This can be used to evaluate improper integrals over intervals unbounded on both sides.

Integrals in which the integrand approaches ∞ or $-\infty$ in the interval:

If f is continuous on the interval (a,b) but $f(x) \to \infty$ or $-\infty$ as $x \to b$ from the left, then we can define

$$\int_{a}^{b} f(x)dx = \lim_{k \to b^-} \int_{a}^{k} f(x)dx, \text{ if the limit exists.}$$

Similarly, if $f(x) \to \infty$ or $-\infty$ as $x \to a$ from the right, then we can define

$$\int_a^b f(x)dx = \lim_{k \to a^+} \int_k^b f(x)dx \text{, if the limit exists.}$$

Example 31

Evaluate $\int_1^3 \dfrac{1}{(x-3)^2} dx$

Solution:

$$\int_1^3 \frac{1}{(x-3)^2} dx = \lim_{k \to 3^-} \int_1^k \frac{1}{(x-3)^2} dx$$

$$= \lim_{k \to 3^-} \left[\frac{-1}{x-3} \right]_1^k$$

$$= \lim_{k \to 3^-} \left[\frac{-1}{k-3} - \frac{1}{2} \right]$$

$$= \infty$$

Thus, the integral diverges.

Example 32

Evaluate $\int_0^2 \dfrac{1}{(1-x)^2} dx$

Solution:

Here, we observe that $f(x) \to \infty$ as $x \to 1$
Therefore, we write

$$\int_0^2 \frac{1}{(1-x)^2} dx = \int_0^1 \frac{1}{(1-x)^2} dx + \int_1^2 \frac{1}{(1-x)^2} dx$$

$$= \lim_{k \to 1^-} \int_0^k \frac{1}{(1-x)^2} dx + \lim_{k \to 1^+} \int_k^2 \frac{1}{(1-x)^2} dx$$

$$= \lim_{k \to 1^-} \left[\frac{1}{1-x} \right]_0^k + \lim_{k \to 1^+} \left[\frac{1}{1-x} \right]_k^2$$

INTEGRALS

$$= \lim_{k \to 1^-}\left[\frac{1}{1-k} - 1\right] + \lim_{k \to 1^+}\left[-1 - \frac{1}{1-k}\right]$$

$$= \infty$$

Therefore, the improper integral is divergent.

Solve the following:

1. $\int_0^\infty e^{-3x} dx$

2. $\int_{-\infty}^\infty x^3 dx$

3. $\int_{-\infty}^2 \frac{1}{x^2+4} dx$

4. $\int_0^2 \frac{1}{\sqrt{x-2}} dx$

5. $\int_0^4 \frac{1}{x-3} dx$

APPLICATIONS OF INTEGRALS

We know that on integrating a function, we obtain an antiderivative of the function. It may represent different functions depending on different values of the constant C.

In this section, we shall see how to obtain a specific antiderivative of the given function based on certain initial conditions.

Consider a function $f(x) = 6x$.

Now, $\int f(x)dx = 3x^2 + C$

For different values of C, the above function $F(x) = 3x^2 + C$ will represent different curves. But, if we now say that we want to find the curve that passes through the point (3, 5) then we have:

$$F(3) = 5 = 3(3)^2 + C$$

$$\Rightarrow C = -22$$

∴ The specific antiderivative is $F(x) = 3x^2 - 22$

Let us now look into a more practical application of integrals. We have studied in applications of derivatives that acceleration is the derivative of velocity, which is the derivative of displacement, with respect to time.

i.e. $a(t) = v'(t)$, and

$v(t) = s'(t)$.

We observe that $v(t)$ is an antiderivative of $a(t)$ and $s(t)$ is an antiderivative of $v(t)$. We use this to find $v(t)$ and $s(t)$, as follows: If the acceleration is constant and equal to a, then

$$v(t) = \int a\,dt$$
$$= at + C$$

If, $v(0) = v_0$ is the initial velocity at time $t = 0$, then

$$v(0) = a(0) + C$$
$$\therefore v_0 = C$$

Thus, we have:

$$v(t) = at + v_0$$

Similarly,

$$s(t) = \int v\,dt$$
$$= \int (at + v_0)\,dt$$
$$= \frac{1}{2}at^2 + v_0 t + C$$

Again, if we take an initial displacement of $s(0) = s_0$ at time $t = 0$, we have

$$s(0) = \frac{1}{2}(a)(0)^2 + (v_0)(0) + C$$
$$\Rightarrow C = s_0$$
$$\therefore s(t) = \frac{1}{2}at^2 + v_0 t + s_0$$

The above equation is called the **equation of motion** of a particle in a straight line with constant acceleration.

Example 33

A ball is thrown straight up from a platform 8 meters above the ground with an initial velocity of 49 m/s. If gravity is the only force acting on the ball, calculate the height the ball reaches before it starts its descent.

INTEGRALS

Solution:

$$a = -g = -9.8 \text{ m/s}$$

(**Note**: The negative sign appears since the force of gravity acts in the opposite direction of motion)

$$v_0 = 49 \text{ m/s}$$
$$s_0 = 8 \text{ m.}$$

Now, the ball will continue to rise untill its velocity becomes zero
i.e. $v(t) = 0$

We have,

$$v(t) = a(t) + v_0$$
$$0 = -gt + 49$$

$$\therefore t = 5 \text{ sec}$$

The height that the ball reaches in $t = 5$ seconds is

$$s(t) = \frac{1}{2}gt^2 + v_0 t + s_0 \quad s(t) = -\frac{1}{2}gt^2 + v_0 t + s_0$$

$$\therefore s(5) = \frac{1}{2}(9.8)(5)^2 + 49(5) + 8$$

$$= 130.5 \text{ m}$$

Solve the following:

1. A man drops a stone from a building top. The stone hits the ground 5 seconds later. Find its velocity at impact and the height of the building. Assume that the stone falls freely under gravity.

2. A car traveling at 60 mi/hr on a straight road starts to decelerate constantly at 10 ft/sec². How long will it take for the car to reduce to a speed of 45 mi/hr?
   How far will the car travel before it comes to a complete stop?

3. A car traveling at 60 mi/hr skids 180 feet after brakes are applied before coming to a stop. Find the deceleration of the car, assuming it to be constant.

Solutions Of Differential Equations

In this section, we shall use integrals in solving differential equations. A differential equation $\frac{dy}{dx} = f(x,y)$ has a solution in terms of relation between x and y of the form $g(x,y) = 0$

A differential equation is **separable** if it can be expressed in the form

$$\frac{dy}{dx} = \frac{h(x)}{g(y)}$$

To solve such a separable differential equation, we first write it in the form $g(y)dy = h(x)dx$

Integrating both sides, we have:

$$\int g(y)dy = \int h(x)dx$$

The above integral can be solved using standard techniques to obtain a solution to the differential equation.

Example 34

Solve the equation $\frac{dy}{dx} = \frac{x}{y^2}$

Solution:

We have:

$$\frac{dy}{dx} = \frac{x}{y^2}$$

Writing the equation in separable form, we get:

$$y^2 dy = x dx$$

Integrating both sides, we have:

$$\int y^2 dy = \int x dx$$

$$\therefore \frac{y^3}{3} = \frac{x^2}{2} + C$$

In explicit form, we have:

INTEGRALS

$$y = (\frac{3}{2}x^2 + 3C)^{1/3}$$

$$y = (\frac{3}{2}x^2 + K)^{1/3}, \text{ where } K = 3C.$$

To obtain the value of K, we may have to use an initial condition given to us, if any. A differential equation with an initial condition is called an **Initial Value Problem**.

Example 35

Solve the initial value problem $\frac{dy}{dx} = 4xy^2$; $y(0) = 2$.

Solution:

Writing the equation in separable form, we have:

$$\frac{dy}{y^2} = 4x\,dx$$

$$\therefore \int \frac{dy}{y^2} = \int 4x\,dx$$

$$\therefore \frac{-1}{y} = 2x^2 + C$$

$$\therefore y = \frac{-1}{2x^2 + C}$$

Now, at $x = 0$, we have: $y = 2$

$$\therefore 2 = \frac{-1}{C}$$

$$\Rightarrow C = -\frac{1}{2}$$

$$\therefore y = \frac{-1}{2x^2 - \frac{1}{2}}$$

$$= \frac{2}{1 - 4x^2}$$

Differential Equations Involving Exponential Growth:
A quantity is said to have exponential growth (or decay) if at each instant of

time, its rate of increase (or decrease) is proportional to the amount of quantity present.

If $y(t)$ is the amount of quantity present at time t, then since the rate of change is proportional to the quantity present, we have:

$$\frac{dy}{dt} \propto y$$

or, $\frac{dy}{dt} = ky$, where k = constant of proportionality.

To solve such an equation, we write the equation in the separable differential form as follows:

$$\frac{dy}{y} = kdt$$

Integrating, we get:

$$\int \frac{dy}{y} = \int kdt$$

$$\ln y = kt + C'$$

$$\Rightarrow y = e^{kt+C'}$$

or, $y = e^{kt} e^{C'}$

By writing $C' = \ln C$ and modifying our equation, we end up with:

$$y = Ce^{kt}$$

Now, at $t = 0$, if $y(0) = y_0$, then we have:

$$y_0 = Ce^0 = C$$

$$y = y_0 e^{kt}$$

The constant "k" is called the **Growth (or Decay) Constant**.

Examples of exponential growth can be found in population growth, radioactive decay, etc.

Example 36

Find the population of a particular country in 25 years at a 2% annual growth rate, if it started at an initial population of 2 million people.

Solution:

In this case, if $y(t)$ = population at time t, then we have:
$y(0) = 2$ million
$k = 0.02$
$t = 25$

$$y(t) = y_0 e^{kt}$$
$$\therefore y(25) = 2e^{0.02(25)}$$
$$= 3.2974 \text{ million.}$$

Solve the following problems:

1. The number of bacteria in a certain culture grows exponentially at the rate of 2% per hour. If 5,000 bacteria are present initially, find:

   a. The number of bacteria present in 3 hours, and
   b. Time required for the population to reach 25,000.

2. A radioactive element has a half-life of 120 days. If a sample of that element initially weighs 20 grams, find the amount remaining in 45 days. (*Hint*: The half-life of a radioactive element is the time taken by that element to reduce to half.)

NUMERICAL APPROXIMATIONS TO DEFINITE INTEGRALS

We have already seen a way to approximate the value of a definite integral by using a Riemann Sum. In this section, we shall look into another method of approximation to definite integrals called the **Trapezoidal Rule.**

Consider a curve $y = f(x)$ as shown below. We partition the interval $[a,b]$ into equal subintervals of length Δx by introducing points $x_1, x_2, ..., x_{n-1}$ between a and b.

We then have: $\Delta x = \dfrac{b-a}{n}$

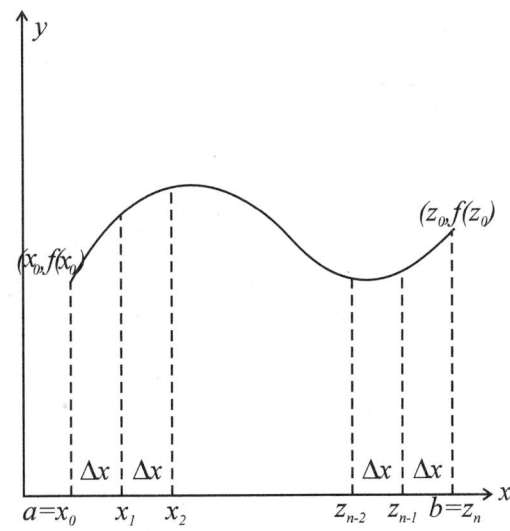

The area under the curve $y = f(x)$ can be approximated as the sum of areas of n trapeziums formed on the previous page. The area of a typical trapezium, shown below, can be calculated as follows:

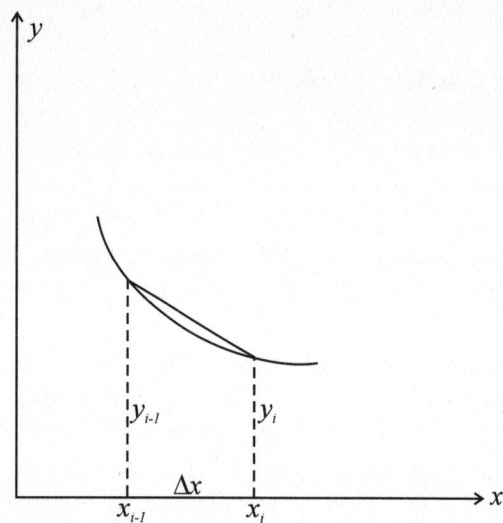

$$A_i = \left[\frac{y_{i-1} + y_i}{2}\right]\Delta x, \text{ where } y_i = f(x_i)$$

Therefore, the net area under the curve is approximately equal to the sum of areas of the n trapeziums. Thus,

$$\int_a^b f(x)dx \approx \sum_{i=1}^n A_i$$

$$= \left[\frac{f(x_0) + f(x_1)}{2}\right]\Delta x + \left[\frac{f(x_1) + f(x_2)}{2}\right]\Delta x + \ldots + \left[\frac{f(x_{n-1}) + f(x_n)}{2}\right]\Delta x$$

$$= \frac{\Delta x}{2}\left[f(x_0) + 2f(x_1) + 2f(x_2) + \ldots + 2f(x_{n-1}) + f(x_n)\right]$$

$$= \frac{b-a}{2n}\left[f(x_0) + 2f(x_1) + 2f(x_2) + \ldots + 2f(x_{n-1}) + f(x_n)\right]$$

Example 37
Approximate $\int_1^2 \frac{1}{x}dx$ using the Trapezoidal Rule. Take $n = 5$ and then $n = 10$.

Solution:

a. $n = 5$

$$\therefore \Delta x = \frac{b-a}{n} = 0.2$$

We have:

$x_0 = a = 1 \Rightarrow f(x_0) = 1$
$x_1 = 1.2 \Rightarrow f(x_1) = 1/1.2$
$x_2 = 1.4 \Rightarrow f(x_2) = 1/1.4$

and so on, finally

$x_5 = b = 2.0 \Rightarrow f(x_5) = 1/2$

$$\therefore \int_1^2 \frac{1}{x} dx = \frac{0.2}{2}\left[f(x_0) + 2f(x_1) + 2f(x_2) + 2f(x_3) + 2f(x_4) + f(x_5)\right]$$

$$= 0.1\left[1 + \frac{2}{1.2} + \frac{2}{1.4} + \frac{2}{1.6} + \frac{2}{1.8} + \frac{1}{2}\right]$$

$$= 0.6956$$

b. $n = 10$

$$\therefore \Delta x = \frac{b-a}{n} = 0.1$$

Therefore,

$x_0 = a = 1 \Rightarrow f(x_0) = 1$
$x_1 = 1.1 \Rightarrow f(x_1) = 1/1.1$
$x_2 = 1.2 \Rightarrow f(x_2) = 1/1.2$

and so on, finally

$x_{10} = b = 2.0 \Rightarrow f(x_{10}) = 1/2$

$$\therefore \int_1^2 \frac{1}{x} dx = \frac{0.1}{2}\left[f(x_0) + 2f(x_1) + 2f(x_2) + \ldots + 2f(x_9) + f(x_{10})\right]$$

$$= 0.1\left[1 + \frac{2}{1.1} + \frac{2}{1.2} + \frac{2}{1.3} + \ldots + \frac{2}{1.9} + \frac{1}{2}\right]$$

$$= 0.6938$$

Note: The actual area using the Fundamental Theorem of Calculus is

$$\int_1^2 \frac{1}{x}dx = \left[\ln|x|\right]_1^2$$
$$= \ln 2 - \ln 1$$
$$= 0.6932$$

Approximate the following using the Trapezoidal Rule:

1. $\int_0^1 \frac{1}{x+1}dx$, $n=4$

2. $\int_1^4 x^2 dx$, $n=6$.
   Compare the result with the exact area obtained using the **Fundamental Theorem of Calculus**.

3. $\int_0^\pi \sin x\, dx$, $n=6$

4. $\int_0^2 e^{x^2} dx$, $n=5$

1. $\int_0^1 \frac{1}{x+1} dx, n=4$

$$n=4 \therefore \frac{b-a}{n} = \frac{1-0}{4} = \frac{1}{4} = 0.25$$

$x_0 = 0 \rightarrow f(x_0) = 1$
$x_1 = .25 \rightarrow f(x_1) = 1/1.25$
$x_2 = .50 \rightarrow f(x_2) = 1/1.50$
$x_3 = .75 \rightarrow f(x_3) = 1/1.75$
$x_4 = 1 \rightarrow f(x_4) = 1/2$

Thus, $\int_0^1 \frac{1}{x+1} dx$ from 0 to 1 $= \frac{.25}{2}[1 + 2/1.25 + 2/1.50 + 2/1.75 + 1/2]$

$$= \frac{.25}{2}[1 + 1.6 + 1.33 + 1.14 + .5]$$
$$= 0.69625$$

2. $\int_1^4 x^2 dx$, $n = 6$

$$n = 6 \therefore \frac{b-a}{n} = \frac{4-1}{6} = \frac{3}{6} = \frac{1}{2} = .5$$

$x_0 = 1 \rightarrow f(x_0) = 1$
$x_1 = 1.5 \rightarrow f(x_1) = 2.25$
$x_2 = 2 \rightarrow f(x_2) = 4$
$x_3 = 2.5 \rightarrow f(x_3) = 6.25$
$x_4 = 3 \rightarrow f(x_4) = 9$
$x_5 = 3.5 \rightarrow f(x_5) = 12.25$
$x_6 = 4 \rightarrow f(x_6) = 16$

Thus, $\int_1^4 x^2 dx = \frac{.5}{2}[1 + 2 \cdot 2.25 + 2 \cdot 4 + 2 \cdot 6.25 + 2 \cdot 9 + 2 \cdot 12.25 + 16]$

$= \frac{.5}{2}[1 + 4.5 + 8 + 12.5 + 18 + 24.5 + 16]$

$= 21.125$

The remaining two problems are left for the student as exercises using this, or a similar format for the solutions.

SLOPE-FIELDS

A method for graphing first-order differential equations exists that results in something called a slope-field. Slope-fields are also known as 'direction-fields.' A slope field is a graph containing a plot of points within a finite domain, each having a tangent line drawn through or adjacent to them. The slope of each tangent line is obtained from the value of the differential equation at that point. Once you have plotted points, calculated slopes, and drawn the tangent lines in the slope-field, a curve can be drawn. This curve is the solution curve for the differential equation.

What we already know about derivatives is that if the derivative of $y = f(x)$ exists for some point (x_i, y_j), then *the slope of the curve at that point is the value of the function $f'(x_i)$*. It is important to remember this fact when plotting solution curves as it will save you considerable time when working with slope-fields.

As an example, let's plot the slope-field for the differential equation $dy/dx = y$. Applying our rule of thumb from above, we know that $dy/dx = y$ will have a solution curve where the slope will be the value of its' y-coordinate at any point (x_i, y_j). Essentially, y is a function of the derivative and $dy/dx = y$ can be reinterpreted as:

Slope = y

To graph the slope field for this differential equation, we first plot a series of evenly spaced points in quadrants I and II (which quadrant(s) you plot points in will be determined by the nature of the problem). At each point, a small line is drawn tangent to the point that has slope equal to the value of the function at that point. Graph A shows the tangent lines at each plotted point on the graph. Note that the value of the slope for each point is the value of the function at that point.

Graph A:

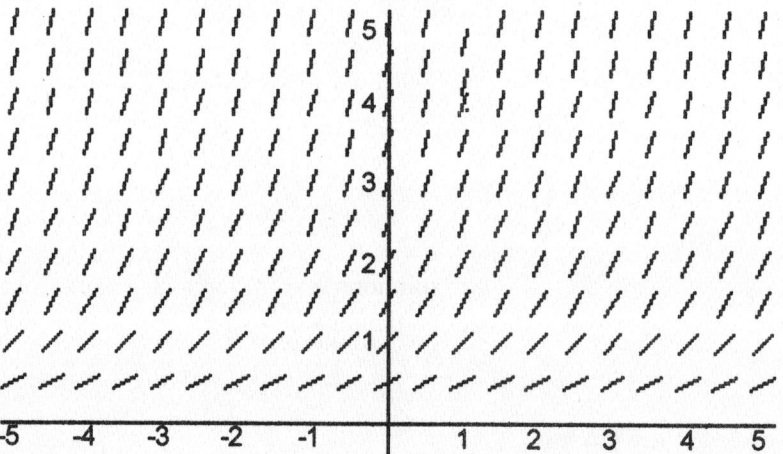

Once the points have been plotted and the tangent lines drawn, a solution curve can be drawn. Typically, in problems involving slope-fields, you are given a value for the independent variable in the differential equation. In this case, we will specify that $x = 0$. At $x = 0$, you see that the slope of the tangent at the point $(0,1)$ is 1. Graph B depicts the solution curve for the differential equation $dy/dx = y$.

Graph B:

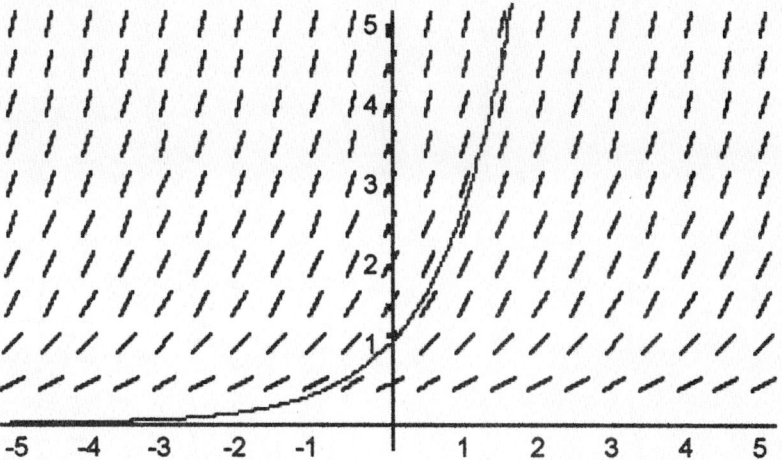

In working with slope-fields and the differential equations that generate them, it's easy to see that this method yields an infinite number of potential solutions for a differential equation. In the AP exam, you are asked for a particular solution.

In arriving at a particular solution, we must first understand that first order *initial value problems* are usually in two parts: a first order differential equation *and* an initial condition. It's useful for the student to bear in mind that slope-fields are often used in the study of physics where the solution graph is referred to as a *trajectory*. Why this is so has to do with the nature of particles accelerating through space. In all cases where problems of this type present themselves (and calculus is used in the solution of virtually all of these problems), the student is told that there exists a starting point for the acceleration. This is usually given as "time = 0" and describes that moment when the body (or particle) is at rest. The value for the variable time, in such a case, represents the *initial condition*. The same is true when solving differential equations using slope-fields.

In the vast majority of cases, you are given an initial condition for the solution curve. The initial condition will be a value assigned to the independent variable in the differential equation. As you will see, most of these values will be either 0 or 1 and represent a starting point for the solution curve or signal a change in the trajectory of a body in motion.

Another example is the graph of the differential equation $dy/dx\ x^2/y$, where $x_0 = 0$. This is shown in C:

Graph C:

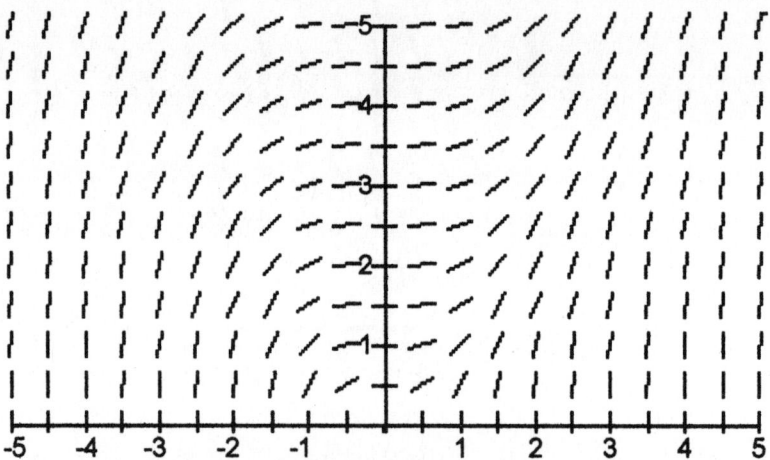

When the solution curve is plotted, it passes through the point indicated in the initial condition. In this case, when $x_0 = 0$, $y = 1$, as shown in graph D.

Graph D:

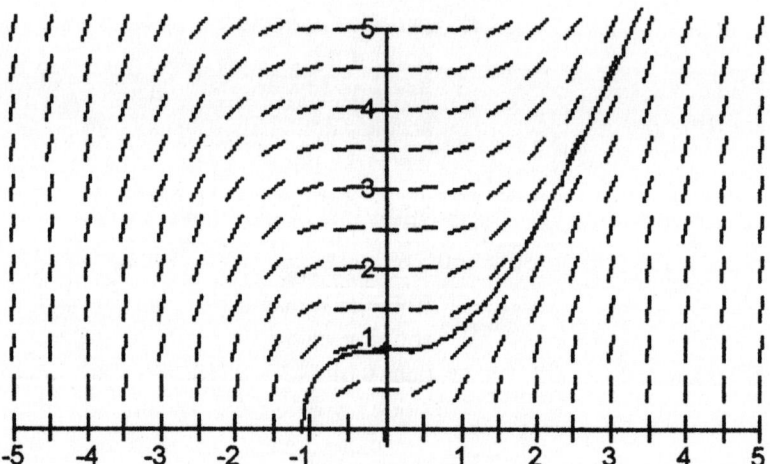

Part IV
SEQUENCES AND SERIES

SEQUENCES OF REAL NUMBERS

A **sequence** $\{a_n\}$ is a function whose domain is the set of positive integers. An arbitrary infinite sequence, usually referred to simply as a sequence, is often denoted as follows:

$$a_1, a_2, a_3, \ldots a_n, \ldots$$

where a_1, denotes the first term of the sequence, a_2, denotes the second term, and a_n, the nth term of the sequence. The three dots at the end indicate that the sequence continues with infinitely many terms; i.e. the sequence does not terminate.

If a sequence is defined by $a_n = \dfrac{n}{n+1}$, then $a_1 = \dfrac{1}{2}, a_2 = \dfrac{2}{3}, a_3 = \dfrac{3}{4}, \ldots$.

In general, the equation $a_n = \dfrac{n}{n+1}$ defines the sequence

$$\dfrac{1}{2}, \dfrac{2}{3}, \dfrac{3}{4}, \ldots, \dfrac{n}{n+1}, \ldots .$$

Example 1
List the first four terms of the sequence defined by $a_n = \dfrac{3n}{2+n}$.

Solution:

$$a_1 = \dfrac{3(1)}{2+1} = \dfrac{3}{3} = 1; \quad a_2 = \dfrac{3(2)}{2+2} = \dfrac{6}{4} = \dfrac{3}{2}; \quad a_3 = \dfrac{3(3)}{2+3} = \dfrac{9}{5}; \quad a_4 = \dfrac{3(4)}{2+4} = \dfrac{12}{6} = 2$$

Hence, the first four terms of the given sequence are $1, \dfrac{3}{2}, \dfrac{9}{5},$ and 2.

Example 2
List the first four terms of the sequence defined by $a_n = 3 + \dfrac{(-1)^{n+1}}{2n}$.

Solution:

$$a_1 = 3 + \dfrac{(-1)^{1+1}}{2(1)} = 3 + \dfrac{1}{2} = \dfrac{7}{2}$$

$$a_2 = 3 + \frac{(-1)^{2+1}}{2(2)} = 3 - \frac{1}{4} = \frac{11}{4}$$

$$a_3 = 3 + \frac{(-1)^{3+1}}{2(3)} = 3 + \frac{1}{6} = \frac{19}{6}$$

$$a_4 = 3 + \frac{(-1)^{4+1}}{2(4)} = 3 - \frac{1}{8} = \frac{23}{8}$$

Hence, the first four terms of the given sequence are $\frac{7}{2}, \frac{11}{4}, \frac{19}{6},$ and $\frac{23}{8}$.

Example 3
List the first three terms of the sequence defined by $a_n = n\sin(n)$.

Solution:

$a_1 = (1)\sin(1) = \sin(1)$
$a_2 = (2)\sin(2) = 2\sin(2)$
$a_3 = (3)\sin(3) = 3\sin(3)$

Hence, the first three terms of the given sequence are sin (1), 2sin(2), and 3sin(3).

Example 4
Determine the fifth and ninth terms of the sequence defined by

$$a_n = 3 + (0.1)^{n-2}$$

Solution:

The fifth term of the sequence is

$$a_5 = 3 + (0.1)^{5-2} = 3 + (0.1)^3 = 3 + 0.001 = 3.001$$

The ninth term of the sequence is

$$a_9 = 3 + (0.1)^{9-2} = 3 + (0.1)^7 = 3 + 0.0000001 = 3.0000001$$

A sequence $\{a_n\}$ has the **limit L**, or **converges** to L, denoted by either $\lim_{n \to \infty} a_n = L$ or $a_n \to L$ as $n \to \infty$

if the difference between a_n and L can be made arbitrarily close to 0 by taking n sufficiently large. If such a number L does not exist, then the sequence has no limit, or **diverges**.

SEQUENCES AND SERIES

A sequence that converges is called a **convergent sequence**. A sequence that diverges is called a **divergent sequence**.

The following are examples of convergent sequences.

Example 5
The sequence defined by $a_n = \dfrac{3}{e^n}$ converges to 0 since
$$\lim_{n\to\infty} a_n = \lim_{n\to\infty} \frac{3}{e^n} = 0$$

Example 6
The sequence defined by $a_n = \dfrac{4n+1}{n^2 - 3n}$ converges to 0 since
$$\lim_{n\to\infty} a_n = \lim_{n\to\infty} \frac{4n+1}{n^2-3n} = \lim_{n\to\infty} \frac{4}{2n-3} = 0. \text{ (Using L'Hôpital's Rule)}$$

Example 7
The sequence defined by $a_n = \dfrac{n^3 + 2n + 1}{3n^3 - 4}$ converges to $\dfrac{1}{3}$ since
$$\lim_{n\to\infty} a_n = \lim_{n\to\infty} \frac{n^3+2n+1}{3n^3-4} = \lim_{n\to\infty} \frac{3n^2+2}{9n^2} = \lim_{n\to\infty} \frac{6n}{18n} = \lim_{n\to\infty} \frac{6}{18} = \frac{1}{3}.$$

(Using L'Hôpital's Rule)

Note that if a sequence has a limit, the limit is unique. The following are examples of **divergent sequences**.

Example 8
The sequence defined by $a_n = (-1)^n$ diverges since
$\{a_n\} = \{(-1)^n\} = \{-1, 1, -1, 1, ...\}$ oscillates between –1 and 1 and, therefore, has no limit.

Example 9
The sequence defined by $a_n = \cos(n)$ diverges since
$\{a_n\} = \{\cos(n)\} = \cos(1), \ \cos(2), \ \cos(3), \ ...$ continuously takes all values in the closed [–1, 1] interval and, therefore, has no limit.

Example 10
The sequence defined by $a_n = \dfrac{e^n}{n^3}$ diverges to infinity since
$$\lim_{n\to\infty} a_n = \lim_{n\to\infty} \frac{e^n}{n^3} = \lim_{n\to\infty} \frac{e^n}{3n^2} = \lim_{n\to\infty} \frac{e^n}{6n} = \lim_{n\to\infty} \frac{e^n}{6} \to \infty.$$

The following definitions are also useful when working with sequences.

Definitions

a. The sequence $\{a_n\}$ is said to be a **nondecreasing** sequence if, for all n, $a_n \leq a_{n+1}$.

b. The sequence $\{a_n\}$ is said to be a **nonincreasing** sequence if for all n, $a_n \geq a_{n+1}$.

c. A sequence $\{a_n\}$ is said to be **monotonic** if it is either non-increasing or nondecreasing.

d. A sequence $\{a_n\}$ is said to be **bounded** if $|a_n| \leq M$ for all n (where M is a real number).

Note that a monotonic sequence converges if and only if it is bounded. An unbounded sequence diverges. Also note that a bounded sequence may diverge. *(See Example 9)*

Series of Constants

Let $\{a_n\}$ be a sequence. We form another sequence $\{S_n\}$ such that

$$S_1 = a_1$$
$$S_2 = a_1 + a_2$$
$$S_3 = a_1 + a_2 + a_3$$
$$\vdots$$
$$S_n = a_1 + a_2 + a_3 + \ldots + a_n$$

The infinite summation $a_1 + a_2 + a_3 + \ldots + a_n + \ldots = \sum_{i=1}^{\infty} a_i$ is called an **infinite series** or simply a **series**. The sequence $\{S_n\}$ is called the **sequence of partial sums**.

The following statements are noted:

1. If $\lim_{n \to \infty} S_n = S$ (where S is finite), then the series $\sum_{i=1}^{\infty} a_i$ **converges** to S and S is called the **sum of the series**.

2. If $\lim_{n \to \infty} S_n$ does not exist, the series $\sum_{i=1}^{\infty} a_i$ **diverges**.

SEQUENCES AND SERIES

We now list some properties of infinite series.
Let k, A, and B be real numbers such that

$$\sum_{i=1}^{\infty} a_i = A, \quad \sum_{i=1}^{\infty} b_i = B, \quad \text{and} \quad S_n = \sum_{i=1}^{n} a_i$$

Then:

1. $\sum_{i=1}^{\infty} k a_i = k \sum_{i=1}^{\infty} a_i = kA$

2. $\sum_{i=1}^{\infty} (a_i \pm b_i) = \sum_{i=1}^{\infty} a_i \pm \sum_{i=1}^{\infty} b_i = A \pm B$

3. $\sum_{i=n+1}^{\infty} a_i = \sum_{i=1}^{\infty} a_i - \sum_{i=1}^{n} a_i = A - S_n$

The third property above suggests that if the first n terms of a series are dropped, then the convergence or divergence of a series is unaffected. However, dropping the first n terms of a convergent series may affect the sum to which the series converges.

In the remainder of the section, we shall try to determine whether a given series converges or diverges. Further, we shall try to determine the sum of a convergent series. The following tests will be useful in our pursuit.

N-TH TERM TEST FOR DIVERGENCE

If the series $\sum_{n=1}^{\infty} a_n$ converges, then $\lim_{n \to \infty} a_n = 0$. Equivalently, we have that if $\lim_{n \to \infty} a_n \neq 0$, then the series **diverges**.

The above test gives a sufficient condition for the divergence of a series. That is, if $\lim_{n \to \infty} a_n = 0$, then the series may converge or diverge. That is, $\lim_{n \to \infty} a_n = 0$ is a **necessary** condition for convergence of the series but it is **not** a sufficient condition.

Example 11
Using the n-th Term Test, determine which, if any, of the following series diverge:

a. $\sum_{n=1}^{\infty} \frac{1}{n}$ b. $\sum_{n=1}^{\infty} e^n$ c. $\sum_{n=1}^{\infty} \sin(n)$ d. $\sum_{n=1}^{\infty} \frac{n^2 + 1}{n^3 - 3}$

Solution:

a. Since $\lim_{n\to\infty} \frac{1}{n} = 0$, we draw no conclusion about convergence or divergence of the given series.

b. Since $\lim_{n\to\infty} e^n \neq 0$, we conclude that the given series diverges.

c. Since $\lim_{n\to\infty} \sin(n) \neq 0$, we conclude then that given series diverges.

d. Since $\lim_{n\to\infty} \frac{n^2+1}{n^3-3} = \lim_{n\to\infty} \frac{2n}{3n^2} = \lim_{n\to\infty} \frac{2}{6n} = 0$, we draw no conclusion about convergence or divergence of the given series.

GEOMETRIC SERIES TEST

The series given by $\sum_{n=0}^{\infty} ar^n = a + ar + ar^2 + \ldots + ar^n + \ldots \quad (a \neq 0)$ is called a geometric series with ratio r.

a. If $r \geq 1$ then the series **diverges**.

b. If $r < 1$, then the series **converges** and has the sum $\sum_{n=0}^{\infty} ar^n = \frac{a}{1-r}$.

Example 12
Using the Geometric Series Test, determine the convergence or divergence of the following series.

a. $1 + 0.7 + 0.49 + 0.343 + 0.2401 + \ldots$
b. $2 + 1.8 + 1.62 + 1.458 + 1.3122 + \ldots$
c. $1 + 1.2 + 1.44 + 1.728 + 2.0736 + \ldots$
d. $3 + 3.3 + 3.63 + 3.993 + 4.3923 + \ldots$

Solution:

a. The geometric series $1 + .07 + 0.49 + 0.343 + 0.2401 + \ldots$ may be written as

$$\sum_{n=0}^{\infty} 0.7^n = 1 + 0.7 + (0.7)^2 + (0.7)^3 + (0.7)^4 + \ldots$$

with $a = 1$ and $r = 0.7$.

Since $r = 0.7 < 1$, the series converges and its sum is $\frac{a}{1-r} = \frac{1}{1-0.7} = \frac{1}{0.3} = \frac{10}{3}$.

SEQUENCES AND SERIES

b. The geometric series $2 + 1.8 + 1.62 + 1.458 + 1.3122 + \ldots$ may be written as $\sum_{n=0}^{\infty} 2(0.9)^n = 2 + 2(0.9) + 2(0.9)^2 + 2(0.9)^3 + 2(0.9)^4 + \ldots$ with $a = 2$ and $r = 0.9$. Since $r = 0.9 < 1$, the series converges and its sum is $\frac{a}{1-r} = \frac{2}{1-0.9} = \frac{2}{0.1} = 20$.

c. The geometric series $1 + 1.2 + 1.44 + 1.728 + 2.0736 + \ldots$ may be written as $\sum_{n=0}^{\infty}(1.2)^n = 1 + (1.2) + (1.2)^2 + (1.3)^3 + (1.2)^4 + \ldots$ with $a = 1$ and $r = 1.2$. Since $r > 1$, the series diverges.

d. The geometric series $3 + 3.3 + 3.63 + 3.993 + 4.3923 + \ldots$ may be written as $\sum_{n=0}^{\infty} 3(1.1)^n = 3 + 3(1.1) + 3(1.1)^2 + 3(1.1)^3 + 3(1.1)^4 + \ldots$ with $a = 3$ and $r = 1.1 > 1$. Therefore, the series diverges.

P-SERIES TEST

The series $\sum_{n=1}^{\infty} \frac{1}{p^n} = \frac{1}{p} + \frac{1}{p^2} + \frac{1}{p^3} + \ldots$ is a geometric series with $a = 1$ and $r = \frac{1}{p}$. The series converges if $p > 1$ and diverge if $0 < p \leq 1$.

If we interchange the roles of p and n we get the new series

$$\sum_{n=1}^{\infty} \frac{1}{n^p} = \frac{1}{1^p} + \frac{1}{2^p} + \frac{1}{3^p} + \frac{1}{4^p} + \ldots,$$ which is called a **p-series**.

Consider the p-series $\sum_{n=1}^{\infty} \frac{1}{n^p} = \frac{1}{1^p} + \frac{1}{2^p} + \frac{1}{3^p} + \frac{1}{4^p} + \ldots$.

1. If $p > 1$, the series **converges**.

2. If $0 < p \leq 1$, then the series **diverges**.

Example 13
Using the p-series test, determine the convergence or divergence of each of the following series.

a. $\sum_{n=1}^{\infty} \frac{1}{n}$ b. $\sum_{n=1}^{\infty} \frac{1}{n^4}$ c. $\sum_{n=1}^{\infty} \frac{1}{\sqrt{n}}$

Solutions:

a. The series $\sum_{n=1}^{\infty}\frac{1}{n} = 1 + \frac{1}{2} + \frac{1}{3} + \ldots + \frac{1}{n} + \ldots$ is a p-series with $p = 1$. Therefore, the series diverges. This series is called the **harmonic series.**

b. The series $\sum_{n=1}^{\infty}\frac{1}{n^4} = 1 + \frac{1}{2^4} + \frac{1}{3^4} + \ldots + \frac{1}{n^4} + \ldots$ is a p-series with $p = 4$. Therefore, the series **converges.**

c. The series $\sum_{n=1}^{\infty}\frac{1}{\sqrt{n}} = 1 + \frac{1}{\sqrt{2}} + \frac{1}{\sqrt{3}} + \frac{1}{\sqrt{4}} + \ldots + \frac{1}{\sqrt{n}} + \ldots$ is a p-series with $p = \frac{1}{2}$. Therefore, the series diverges.

INTEGRAL TEST

Let f be a continuous function. If f is positive valued and decreasing for all $x \geq 1$, and if $a_n = f(n)$, then

$$\sum_{n=1}^{\infty} a_n \text{ and } \int_{1}^{\infty} f(x)dx \text{ either both converge or both diverge.}$$

Example 14
Show that the p-series $\sum_{n=1}^{\infty}\frac{1}{n^p}$ converges if $p > 1$ and diverges if $0 < p \leq 1$.

Solution:

Let $f(x) = \frac{1}{x^p}$ and consider:

a. If $p > 1$, then $\int_{1}^{\infty}\frac{1}{x^p}\,dx = \int_{1}^{\infty} x^{-p}\,dx = \lim_{k \to \infty}\left[\frac{x^{1-p}}{1-p}\right]_{1}^{k} =$

$\lim_{k \to \infty}\frac{1}{1-p}\left(k^{1-p} - 1\right) = \frac{1}{1-p}\lim_{k \to \infty}\left(k^{1-p} - 1\right) = \frac{1}{1-p}(-1) = \frac{1}{p-1}$ and,

therefore, $\sum_{n=1}^{\infty} n^p$ converges if $p > 1$.

SEQUENCES AND SERIES

b. If $p = 1$, then $\int_1^\infty \frac{1}{x^p} dx$ becomes $\int_1^\infty \frac{1}{x} dx = \lim_{k \to \infty} \ln(k) \to \infty$

and, therefore, $\sum_{n=1}^\infty \frac{1}{n}$ diverges.

c. If $p < 1$, then $\int_1^\infty \frac{1}{x^p} dx = \int_1^\infty x^{-p} dx = \lim_{k \to \infty} \left[\frac{1}{1-p}(x^{1-p}) \right]_1^k =$

$\lim_{k \to \infty} \frac{1}{1-p}(k^{1-p} - 1) \to \infty$ and, therefore, $\sum_{n=1}^\infty \frac{1}{n^p}$ diverges if $p < 1$.

Example 15
Using the Integral Test, determine the convergence or divergence of the

series $\sum_{n=1}^\infty \frac{2n}{n^2 + 1}$.

Solution:

Consider the associated improper integral $\int_1^\infty \frac{2x}{x^2 + 1} dx = \lim_{k \to \infty} \left[\ln(x^2 + 1) \right]_1^k =$

$\lim_{k \to \infty} \left[\ln(k^2 + 1) - \ln(2) \right] \to \infty$.

Therefore, both the improper integral and the series diverge.

Example 16
Using the Integral Test, determine the convergence or divergence of the

series $\sum_{n=1}^\infty \frac{n}{e^n}$.

Solution:

Consider the associated improper integral $\int_1^\infty \frac{x}{e^x} dx = \lim_{k \to \infty} \left[-xe^{-x} - e^{-x} \right]_1^k$

$= \lim_{k \to \infty} \left[(-ke^{-k} - e^{-k}) - \left(-\frac{1}{e} - \frac{1}{e} \right) \right] = 0 + \frac{2}{e} = \frac{2}{e}$ and, therefore, the

given series converges.

The Comparison Test

Let $\sum_{n=1}^{\infty} a_n$ be a series known to either converge or diverge and let $\sum_{n=1}^{\infty} b_n$ be a series under investigation. (Note that a_n and b_n must be positive for all n.)

a. If $\sum_{n=1}^{\infty} a_n$ converges and $b_n \leq a_n$ for all n, then $\sum_{n=1}^{\infty} b_n$ also converges.

b. If $\sum_{n=1}^{\infty} a_n$ diverges and $b_n \geq a_n$ for all n, then $\sum_{n=1}^{\infty} b_n$ also diverges.

The p-series and geometric series are very useful when applying the Comparison Test. The Comparison Test is easy to use but is limited in use dependent upon the knowledge of basic series that are known to either converge or diverge.

Example 17
Using the Comparison Test, determine the convergence or divergence of the series $\sum_{n=1}^{\infty} \frac{1}{3+4^n}$.

Solution:
The given series can be compared to the known **convergent** geometric series

$$\sum_{n=1}^{\infty} \frac{1}{4^n} = \frac{1}{4} + \frac{1}{4^2} + \frac{1}{4^3} + \ldots + \frac{1}{4^n} + \ldots$$

with $a = \frac{1}{4}$ and $r = \frac{1}{4}$. Let $a_n = \frac{1}{4^n}$ and $b_n = \frac{1}{3+4^n}$. Since $a_n = \frac{1}{4^n} > \frac{1}{3+4^n} = b_n$, then the series $\sum_{n=1}^{\infty} \frac{1}{3+4^n}$ also converges.

Example 18
Using the Comparison Test, determine the convergence or divergence of the series $\sum_{n=1}^{\infty} \frac{1}{6+\sqrt{n}}$.

SEQUENCES AND SERIES

Solution:

The given series can be compared with the known **divergent** harmonic series $\sum_{n=1}^{\infty} \frac{1}{n} = 1 + \frac{1}{2} + \frac{1}{3} + \ldots + \frac{1}{n} + \ldots$. Let $a_n = \frac{1}{n}$ and $b_n = \frac{1}{6+\sqrt{n}}$.

Since: $a_n - b_n = \frac{1}{n} - \frac{1}{6+\sqrt{n}} = \frac{6+\sqrt{n}-n}{n(6+\sqrt{n})} \leq 0$ for all $n \leq 9$,

then $a_n \leq b_n$ for all $n \geq 9$.

Therefore, the given series also diverges.

LIMIT COMPARISON TEST

There are times when a given series closely resembles a p-series or a geometric series, yet it is difficult to apply the Comparison Test to test for convergence or divergence. A second comparison test, called the **Limit Comparison Test**, may be useful.

Suppose that a_n and b_n are both positive and that $\lim_{n \to \infty} \frac{a_n}{b_n} = L > 0$. Then the two series $\sum_{n=1}^{\infty} a_n$ and $\sum_{n=1}^{\infty} b_n$ either both converge or both diverge.

Example 19

Using the Limit Comparison Test, determine the convergence or divergence of the series $\sum_{n=1}^{\infty} \frac{1}{4n^2 + 5}$.

Solution:

We choose the **convergent** p-series $\sum_{n=1}^{\infty} \frac{1}{n^2}$ with p = 2. Let

$a_n = \frac{1}{4n^2+5}$ and $b_n = \frac{1}{n^2}$.

Then $\lim_{n \to \infty} \frac{a_n}{b_n} = \lim_{n \to \infty} \left(\frac{1}{4n^2+5} \cdot \frac{n^2}{1} \right) = \lim_{n \to \infty} \frac{n^2}{4n^2+5} = \lim_{n \to \infty} \frac{2n}{8n} =$

$\lim_{n \to \infty} \frac{2}{8} = \frac{1}{4} = L > 0$.

Therefore, the series $\sum_{n=1}^{\infty} \frac{1}{4n^2+5}$ also converges.

Example 20
Using the Limit Comparison Test, determine the convergence or divergence of the series $\sum_{n=1}^{\infty} \dfrac{1}{\sqrt[3]{n^2+4}}$.

Solution:

We choose the **divergent** p-series $\sum_{n=1}^{\infty} \dfrac{1}{\sqrt[3]{n^2}}$ with $p = \dfrac{2}{3}$. Let

$$a_n = \frac{1}{\sqrt[3]{n^2+4}} \quad \text{and} \quad b_n = \frac{1}{\sqrt[3]{n^2}}. \quad \text{Then}$$

$$\lim_{n\to\infty} \frac{a_n}{b_n} = \lim_{n\to\infty} \left(\frac{1}{\sqrt[3]{n^2+4}} \cdot \frac{\sqrt[3]{n^2}}{1} \right) = \lim_{n\to\infty} \frac{\sqrt[3]{n^2}}{\sqrt[3]{n^2+4}} =$$

$$\lim_{n\to\infty} \sqrt[3]{\frac{n^2}{n^2+4}} = \lim_{n\to\infty} \sqrt[3]{\frac{1}{1+4/n^2}} = 1 = L > 0. \text{ Therefore, the series}$$

$\sum_{n=1}^{\infty} \dfrac{1}{\sqrt[3]{n^2+4}}$ diverges.

Alternating Series Test

An **alternating series** is a series whose terms alternate in sign. Hence, the series

$$a_1 - a_2 + a_3 - a_4 + \ldots \quad (a_n > 0) \text{ and}$$

$$-b_1 + b_2 - b_3 + b_4 - \ldots \quad (b_n > 0) \text{ are alternating series.}$$

The alternating series $\sum_{n=1}^{\infty} (-1)^{n-1} a_n = a_1 - a_2 + a_3 - \ldots$ converges, provided that

1. $0 < a_{n+1} \leq a_n$ for all $n \geq 1$, and

2. $\lim_{n\to\infty} a_n = 0$.

Example 21

Determine whether the series $\sum_{n=1}^{\infty} \dfrac{(-1)^n n}{\ln(3n)}$ converges or diverges.

Solution:

The given series is an alternating series. But, $\lim_{n\to\infty} a_n = \lim_{n\to\infty} \frac{n}{\ln(3n)} = \lim_{n\to\infty} \frac{1}{1/n} = \lim_{n\to\infty} n \to \infty \neq 0$. Therefore, the given series diverges by the n-Term Test for Divergence. (Note that the n-th Term Test is valid even for alternating series.)

Example 22

Determine whether the series $\sum_{n=1}^{\infty} (-1)^n \frac{1}{n}$ converges or diverges.

Solution:

The given series is the **alternating harmonic series**.

Since $a_{n+1} = \frac{1}{n+1} \leq \frac{1}{n} = a_n$ for all $n \leq 1$, we have that $a_{n+1} < a_n$. Further, $\lim_{n\to\infty} a_n = \lim_{n\to\infty} \frac{1}{n} = 0$. Therefore, the given alternating series converges.

Definitions

1. A series $\sum_{n=1}^{\infty} a_n$ for which $\sum_{n=1}^{\infty} |a_n|$ converges is called an **absolutely convergent series**. It should be noted that if the series $\sum_{n=1}^{\infty} |a_n|$ converges, then the series $\sum_{n=1}^{\infty} a_n$ also converges.

2. A series which converges but does not converge absolutely is called a **conditionally convergent series**. (The alternating harmonic series converges but the harmonic series does not.)
   (See Examples 13a and 22.)

Alternating Series Remainder

If $\sum_{n=1}^{\infty} (-1)^{n-1} a_n$ is a convergent alternating series with sum S, then the numerical difference between the sum of the first n terms, S_n, and S is less than a_{n+1}. That is, $|S - S_n| < a_{n+1}$.

For a convergent alternating series, its sum S can be approximated by the partial sum S_n. That is, if we stop with the first n terms of the series, then the difference between S and S_n will be less than the first omitted term, a_{n+1}.

Example 23

The alternating series $\sum_{n=1}^{\infty} \frac{(-1)^{n+1}}{n}$ is a convergent series. (*See Example 22*.)

The sum of this series differs from the sum

$$S_7 = 1 - \frac{1}{2} + \frac{1}{3} - \frac{1}{4} + \frac{1}{5} - \frac{1}{6} + \frac{1}{7} \text{ by less than } \frac{1}{8} = a_8,$$

which is called the **error bound**.

Example 24

Determine the error bound of the convergent alternating series

$$\sum_{n=1}^{\infty} (-1)^{n+1} \frac{1}{n!} = \frac{1}{1!} - \frac{1}{2!} + \frac{1}{3!} - \frac{1}{4!} + \frac{1}{5!} - \ldots,$$

if its sum is approximated by S_7.

Solution:

$$S_7 = 1 - \frac{1}{2} + \frac{1}{6} - \frac{1}{24} + \frac{1}{120} - \frac{1}{720} + \frac{1}{5040} \approx 0.63214$$

$$|S - S_7| \leq a_8 = \frac{1}{8!} = \frac{1}{40320} \approx 0.0000248.$$

(The actual sum of this series is $\frac{e-1}{e} \approx 0.63212$.)

Ratio Test

Consider the series $\sum_{n=1}^{\infty} a_n$ such that $\lim_{n \to \infty} \frac{a_{n+1}}{a_n} = L$, if it exists. Then the series converges if $L < 1$, and diverges if $L > 1$.

If $L = 1$, the **Ratio Test** fails; that is, we cannot determine the convergence or divergence of the series. A different test must be used.

Example 25

If possible, use the Ratio Test to determine whether the series $\sum_{n=1}^{\infty} \frac{3^n}{n!}$ converges or diverges.

SEQUENCES AND SERIES

Solution:
Using the Ratio Test, we have

$$\lim_{n\to\infty} \frac{a_{n+1}}{a_n} = \lim_{n\to\infty}\left[\frac{3^{n+1}}{(n+1)!} \cdot \frac{n!}{3^n}\right] = \lim_{n\to\infty}\frac{3}{n+1} = 0 = L.$$

Since $L < 1$, the given series converges.

Example 26

If possible, use the Ratio Test to determine whether the series $\sum_{n=1}^{\infty} \frac{n^n}{2n!}$ converges or diverges.

Solution:
Using the Ratio Test, we have

$$\lim_{n\to\infty}\frac{a_{n+1}}{a_n} = \lim_{n\to\infty}\left[\frac{(n+1)^{n+1}}{2(n+1)!} \cdot \frac{2n!}{n^n}\right] = \lim_{n\to\infty}\frac{(n+1)^{n+1}}{(n+1)\,n^n} = \lim_{n\to\infty}\frac{(n+1)^n}{n^n} =$$

$$\lim_{n\to\infty}\left(1+\frac{1}{n}\right)^n = e = L.$$

Since $L > 1$, the given series diverges.

Example 27

If possible, use the Ratio Test to determine whether the p-series $\sum_{n=1}^{\infty} \frac{1}{n^p}$ converges or diverges.

Solution:
Using the Ratio Test, we have

$$\lim_{n\to\infty}\frac{a_{n+1}}{a_n} = \lim_{n\to\infty}\left[\frac{1}{(n+1)^p} \cdot \frac{n^p}{1}\right] = \lim_{n\to\infty}\frac{n^p}{(n+1)^p} = \lim_{n\to\infty}\left(\frac{1}{1+1/n}\right)^p = 1 = L$$

for all p. Since $L = 1$, the Ratio Test fails. However, for the p-series, we previously determined that the series converges if $p > 1$ and diverges if $p \leq 1$.

Example 28

Determine whether the series $\sum_{n=1}^{\infty}(-1)^n \frac{n^2+1}{3^n}$ is absolutely convergent, conditionally convergent, or divergent.

Solution:

Using the Ratio Test with $\lim\limits_{n\to\infty} \dfrac{a_{n+1}}{a_n}$, we have

$$\lim_{n\to\infty} \frac{a_{n+1}}{a_n} =$$

$$\lim_{n\to\infty}\left[\frac{(n+1)^2+1}{3^{n+1}} \cdot \frac{3^n}{n^2+1}\right] = \frac{1}{3}\lim_{n\to\infty}\frac{(n+1)^2+1}{n^2+1} = \frac{1}{3}\lim_{n\to\infty}\frac{n^2+2n+2}{n^2+1} = \frac{1}{3}(1) = \frac{1}{3} < 1$$

Therefore, the given series is absolutely convergent.

COMPUTATIONS WITH SERIES OF CONSTANTS

Let $\sum\limits_{n=1}^{\infty} a_n$ be a convergent series of positive terms whose sum is S. Then the *nth* partial sum of the series, denoted by $S_n = \sum\limits_{i=1}^{n} a_i$, can be used to approximate S. The error, called the remainder **after n terms**, is denoted by

$$R_n = a_{n+1} + a_{n+2} + a_{n+3} + \ldots \;.$$

Now, consider the positive series $\sum\limits_{n=1}^{\infty} a_n$ that converges by the Integral Test. Then the associated improper integral $\int_1^{\infty} f(x)\, dx$ also converges. We determine that the remainder, R_n, is given by $R_n < \int_n^{\infty} f(x)\, dx$.

The function f is a continuous, positive valued, and decreasing function such that $f(n) = a_n$. In the figure below, the base of each rectangle is one unit and, therefore, the rectangles, starting from the left, have area of $a_2, a_3, a_4, \ldots, a_n, a_{n+1}, \ldots$ and we determine that

$$R_n = a_{n+1} + a_{n+2} + a_{n+3} + \ldots < \int_n^{\infty} f(x)\, dx\,.$$

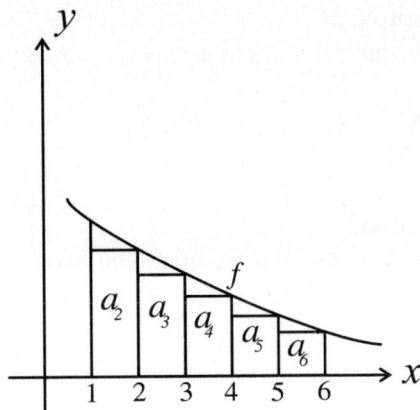

Example 29

Determine the error made when the sum of the series $\sum_{n=1}^{\infty} \frac{2n}{e^{2n}}$ is approximated by its first eight terms.

Solution:

Using the Integral Test, it can be shown that the given series converges. Hence,

$$R_8 < \int_8^{\infty} \frac{2x}{e^{2x}} dx = \lim_{k \to \infty} \left[-xe^{-2x} - \frac{1}{2} e^{-2x} \right]_8^k = \lim_{k \to \infty} \left[-e^{-2x} \left(x + \frac{1}{2} \right) \right]_8^k =$$

$$\lim_{k \to \infty} \left[\frac{k + 1/2}{-e^{2k}} + \frac{17/2}{e^{16}} \right] = \frac{17}{2e^{16}} < 0.00000096 \, .$$

For the convergent, positive term geometric series $\sum_{n=0}^{\infty} ar^n$ with $r < 1$, the remainder after the *nth* term is given by $R_n = \frac{ar^n}{1-r}$.

Example 30

Determine the error made when the sum of the series $\sum_{n=0}^{\infty} \frac{2}{5^n}$ is approximated by its first three terms.

Solution:
Since the given series is a convergent geometric series with $a = 2$ and $r = $, we have

$$R_3 = \frac{ar^3}{1-r} = \frac{2(1/5)^3}{1 - 1/5} = \frac{2(1/125)}{4/5} = \frac{2}{125} \cdot \frac{5}{4} = 0.02 \, .$$

Example 31

Determine how many terms are needed for the error to the sum of the series

$$\sum_{n=0}^{\infty} \frac{2}{5^n}$$ to be less than 0.0002.

Solution:

The given series is a convergent geometric series with $a = 2$ and $r = 1/5$. We must determine the value of n such that

$$R_n = \frac{ar^n}{1-r} < 0.0002 .$$

Hence,

$$\frac{2(1/5)^n}{1 - 1/5} < 0.0002$$

$$2(1/5)^n < 0.00016$$

$$\frac{1}{0.00008} < 5^n$$

$$12,5000 < 5^n$$

$$\ln(12,5000) < n \ln(5)$$

$$n > \frac{\ln(12,5000)}{\ln(5)} \approx 5.86$$

Hence, six terms are required for the stated accuracy.

Power Series

Up to this point, we have been working with series whose terms are constants. We also have series whose terms are variable.

Definitions

1. If x is a variable and all the a_i's are constants, then a series of the form

$$\sum_{i=0}^{\infty} a_i x^i = a_0 + a_1 x + a_2 x^2 + \ldots + a_n x^n + \ldots$$

is called a **power series** in the variable x.

2. A series of the form $\sum_{i=0}^{\infty} a_i (x-c)^i$ is called a **power series in $(x - c)$** or a power series in x centered at c.

SEQUENCES AND SERIES

3. A power series as given prior can be considered as a **function in the variable x**, or $f(x) = \sum_{i=0}^{\infty} a_i(x-c)^i$

   where the domain of f is the set of all x for which the power series converges.

4. The set of x for which a power series converges is called its **interval of convergence**.

Clearly, every power series converges at c, since if $x = c$, then

$$f(c) = \sum_{i=0}^{\infty} a_i(c-c)^i = a_0(1) + a_2(0) + \ldots + a_n(0) + \ldots = a_0.$$

The Ratio Test is used to determine the interval of absolute convergence but, instead of using $\lim_{n\to\infty} \left|\frac{a_{n+1}}{a_n}\right|$, we use $\lim_{n\to\infty}\left|\frac{u_{n+1}}{u_n}\right|$, where $u_i = a_i(x-c)^i$.

Example 32

Determine the interval of convergence for the power series $\sum_{i=0}^{\infty} \frac{(-1)^i x^{2i+1}}{(2i+1)!}$.

Solution:

Let $u_n = \frac{(-1)^n x^{2n+1}}{(2n+1)!}$. Then, $\lim_{n\to\infty}\left|\frac{u_{n+1}}{u_n}\right| =$

$\lim_{n\to\infty}\left|\frac{(-1)^{n+1} x^{2n+3}}{(2n+3)!} \cdot \frac{(2n+1)!}{(-1)^n x^{2n+1}}\right| = \lim_{n\to\infty}\left|\frac{x^{2n+3}}{x^{2n+1}} \cdot \frac{(2n+1)!}{(2n+3)!}\right| =$

$\lim_{n\to\infty}\left|\frac{x^2}{(2n+2)(2n+3)}\right| = |x^2|\lim_{n\to\infty}\left|\frac{1}{(2n+2)(2n+3)}\right| = |x^2|(0) = 0 = L < 1.$

Therefore, by the Ratio Test, the given series converges for all x and its interval of convergence is $(-\infty, \infty)$.

Example 33

Determine the interval of convergence for the power series $\sum_{i=0}^{\infty} i!\, x^i$.

Solution:

Let $u^n = n!\, x^n$. Then,

$\lim_{n\to\infty}\left|\frac{u_{n+1}}{u_n}\right| = \lim_{n\to\infty}\left|\frac{(n+1)!\, x^{n+1}}{n!\, x^n}\right| = \lim_{n\to\infty}\left|(n+1)x\right| = |x|\cdot\lim_{n\to\infty}(n+1) \to \infty.$

Example 34

Determine the interval of convergence for the power series $\sum_{i=0}^{\infty} 2(x-3)^i$.

Solution:

Let $u_n = 2(x-3)^n$. Then,

$$\lim_{n\to\infty} \left| \frac{u_{n+1}}{u_n} \right| = \lim_{n\to\infty} \left| \frac{2(x-3)^{n+1}}{2(x-3)^n} \right| = \lim_{n\to\infty} |x-3| = L.$$

For convergence, $L < 1$. Hence,

$|x-3| < 1$
$-1 < x - 3 < 1$
$2 < x < 4$

Since the Ratio Test fails if $L = 1$, we test the endpoints separately.

If $x = 2$, then $\sum_{i=0}^{\infty} 2(x-3)^i$ becomes

$$\sum_{i=0}^{\infty} 2(2-3)^i = \sum_{i=0}^{\infty} 2(-1)^i = 2 - 2 + 2 - 2 + \ldots,$$ which is

a divergent alternating series.

If $x = 4$, then $\sum_{i=0}^{\infty} 2(x-3)^i$ becomes $\sum_{i=0}^{\infty} 2(4-3)^i = \sum 2(1)^i = 2 + 2 + 2 + 2 + \ldots$ which also diverges.

Hence, the interval of convergence for the given series is $(2, 4)$, which is centered at $c = 3$.

Example 35

Determine the interval of convergence for the power series

$$\sum_{0}^{\infty} \frac{(-1)^i (x+1)^i}{3^i}.$$

Solution:

Let $u_n = \frac{(-1)^n (x+1)^n}{3^n}$. Then,

SEQUENCES AND SERIES

$$\lim_{n\to\infty} \left| \frac{u_{n+1}}{u_n} \right| = \lim_{n\to\infty} \left| \frac{(-1)^{n+1}(x+1)^{n+1}}{3^{n+1}} \cdot \frac{3^n}{(-1)^n(x+1)^n} \right| =$$

$$\lim_{n\to\infty} \left| \frac{(x+1)^{n+1}}{(x+1)^n} \cdot \frac{3^n}{3^{n+1}} \right| = \lim_{n\to\infty} \left| \frac{x+1}{3} \right| = \left| \frac{x+1}{3} \right| = L.$$

For convergence, $L < 1$. Hence,

$$\left| \frac{x+1}{3} \right| < 1$$

$$|x+1| < 3$$

$$-3 < x+1 < 3$$

$$-4 < x < 2$$

We now test at the endpoints.

For $x = -4$, we have

$$\sum_{i=0}^{\infty} \frac{(-1)^i(-4+1)^i}{3^i} = \sum_{i=0}^{\infty} \frac{(-1)^i(-3)^i}{3^i} = \sum_{i=0}^{\infty} (-1)^i(-1)^i =$$

$$\sum_{i=0}^{\infty} (-1)^{2i} = 1 + 1 + 1 + 1 + \ldots \text{ which is a divergent series.}$$

For $x = 2$, we have

$$\sum_{i=0}^{\infty} \frac{(-1)^i(2+1)^i}{3^i} = \sum_{i=0}^{\infty} \frac{(-1)^i 3^i}{3^i} = \sum_{i=0}^{\infty} (-1)^i = 1 - 1 + 1 - 1 + \ldots$$

which is a divergent alternating series.

Thus, the interval of convergence for the given series is $(-4, 2)$, which is centered at $c = -1$.

Example 36

Determine the interval of convergence for the power series $\sum_{i=1}^{\infty} \frac{x^i}{i}$.

Solution:

Let $u_n = \frac{x^n}{n}$. Then,

$$\lim_{n\to\infty}\left|\frac{u_{n+1}}{u_n}\right| = \lim_{n\to\infty}\left|\frac{x^{n+1}}{n+1} \cdot \frac{n}{x^n}\right| = \lim_{n\to\infty}\left|x\frac{n}{n+1}\right| =$$

$$|x|\lim_{n\to\infty}\left|\frac{n}{n+1}\right| = |x|(1) =$$

$$|x| = L.$$

For convergence, $L < 1$. Hence,

$$|x| < 1$$

$$-1 < x < 1$$

We now test for convergence at the endpoints.

For $x = -1$, we have $\sum_{i=1}^{\infty}\frac{(-1)^i}{i} = -1 + \frac{1}{2} - \frac{1}{3} + \frac{1}{4} - \ldots$ which converges by the Alternating Series Test.

For $x = 1$, we have $\sum_{i=1}^{\infty}\frac{(1)^i}{i} = 1 + \frac{1}{2} + \frac{1}{3} + \frac{1}{4} + \ldots$ which is the divergent harmonic series.

Therefore, the interval of convergence for the given series is $[-1, 1)$, which is centered at $c = 0$.

Now, consider the power series $\sum_{i=0}^{\infty}a_i(x-c)^i$ and let $\lim_{n\to\infty}\left|\frac{a_{n+1}}{a_n}\right| = L$ where $0 \leq L < \infty$. Then, the **radius of convergence** of the given series is given by $R = \frac{1}{L}$ where $R = 0$ if $L = \infty$ and $R = \infty$ if $L = 0$.

For each of the intervals of convergence $[-r, r)$, $(-r, r)$, $(-r, r]$, and $[-r, r]$, the radius of convergence is r.

Example 37

Determine the radius of convergence for the power series $\sum_{i=1}^{n}\frac{x^i}{i}$

Solution:

Let $a_n = \frac{1}{n}$. Then,

SEQUENCES AND SERIES

$$\lim_{n \to \infty} \left| \frac{a_{n+1}}{a_n} \right| = \lim_{n \to \infty} \left| \frac{1}{n+1} \cdot \frac{n}{1} \right| = \lim_{n \to \infty} \left| \frac{n}{n+1} \right| = 1.$$

Hence, the radius of convergence is 1. In Example 36, we determined that the interval of convergence for this series is $[-1,1)$, which does have a radius of 1.

POWER SERIES FOR FUNCTIONS

A function defined by a power series behaves like a polynomial. Let f be the function defined by

$$f(x) = \sum_{i=0}^{\infty} a_i(x-c)^i = a_0 + a_1(x-c) + a_2(x-c)^2 + \ldots + a_n(x-c)^n + \ldots (*)$$

whose domain is its interval of convergence. We have the following conclusions:

1. If x_0 is in the domain of f, then $f(x_o) = \sum_{i=0}^{\infty} a_i(x_o - c)^i$.

2. f is continuous for each x in its interval of convergence.

3. The series formed by differentiating each term in the series (*) converges to $f'(x)$ for each x in the domain of f. That is,

   $(**) f'(x) =$

   $$\sum_{i=1}^{\infty} ia_i(x-c)^{i-1} = a_1 + 2a_2(x-c) + 3a_3(x-c)^2 + \ldots + na_n(x-c)^{n-1} + \ldots$$

   We conclude that both the power series (*) and (**) have the same **radius** of convergence but not necessarily the same interval of convergence.

4. The series formed by integrating each term in the series (*) converges to $\sum_{i=0}^{\infty} \frac{a_i(x-c)^{i+1}}{i+1} = a_0(x-c) + \frac{a_1(x-c)^2}{2} + \frac{a_2(x-c)^3}{3} + \ldots + \frac{a_n(x-c)^{n+1}}{n+1} + \ldots$
   for each x in the interval of convergence for the series (*).

Example 38

Let f be the function defined by $f(x) =$

$$\sum_{i=1}^{\infty} \frac{x^i}{i} = x + \frac{x^2}{2} + \frac{x^3}{3} + \dots + \frac{x^n}{n} + \dots .$$ Determine the interval of convergence for both the function f and f'.

Solution:

From Example 37, we determine that the interval of convergence for f is the interval $[-1,1)$. Now,

$$f'(x) = \sum_{i=1}^{\infty} x^{i-1} = 1 + x + x^2 + x^3 + \dots + x^n + \dots .$$

Using the Ratio Test with $u_n = x^{n-1}$, we have

$$\lim_{n \to \infty} \left| \frac{u_{n+1}}{u_n} \right| = \lim_{n \to \infty} \left| \frac{x^n}{x^{n-1}} \right| = \lim_{n \to \infty} |x| = |x| = L.$$

For convergence, $L < 1$. Hence,

$|x| < 1$
$-1 < x < 1$

We now test at the endpoints.

When $x = -1$, we have $\sum_{i=1}^{\infty} (-1)^{i-1} = 1 - 1 + 1 - 1 + \dots$ which is a divergent alternating series.

When $x = 1$, we have $\sum_{i=1}^{\infty} (1)^{i-1} = 1 + 1 + 1 + 1 + \dots$ which also diverges.

Hence, the interval of convergence for f' is $x(-1, 1)$. Note that the series for f and f' have the same radius of convergence but different intervals of convergence.

Example 39

Let f be the function defined by $f(x) = \sum_{i=1}^{\infty} \frac{(-1)^{i+1} x^i}{4^i}$. Determine the interval of convergence for both f and $\int f(x)dx$.

SEQUENCES AND SERIES

Solution:

For the convergence of f, we use the Ratio Test with $u_n = \dfrac{(-1)^{n+1} x^n}{4^n}$.

Hence,

$$\lim_{n\to\infty}\left|\frac{u_{n+1}}{u_n}\right| = \lim_{n\to\infty}\left|\frac{(-1)^{n+2} x^{n+1}}{4^{n+1}} \cdot \frac{4^n}{(-1)^{n+1} x^n}\right| = \lim_{n\to\infty}\left|\frac{x}{4}\right| = L.$$

$$\left|\frac{x}{4}\right| < 1$$

$$|x| < 4$$

$$-4 < x < 4$$

We now test for convergence at the endpoints.

When $x = -4$, we have

$$\sum_{i=1}^{\infty} \frac{(-1)^{i+1}(-4)^i}{4^i} = \sum_{i=1}^{\infty} (-1)^{2i+1} = -1+1-1+1-\ldots \text{ which diverges.}$$

When $x = 4$, we have

$$\sum_{i=1}^{\infty} \frac{(-1)^{i+1}(4)^i}{4^i} = \sum_{i=1}^{\infty} (-1)^{i+1} = 1-1+1-1+\ldots \text{ which also diverges.}$$

Therefore, the interval of convergence for f is $(-4, 4)$.

Now, consider $\displaystyle\int f(x)\,dx = \int \sum_{i=1}^{\infty} \frac{(-1)^{i+1} x^i}{4^i}\,dx = C + \sum_{i=1}^{\infty} \frac{(-1)^{i+1} x^{i+1}}{(i+1)(4)^i} =$

$$C + \frac{x^2}{2(4)} - \frac{x^3}{(3)(4)^2} + \frac{x^4}{(4)(4)^3} - \ldots\,.$$

Using the Ratio Test with $u_n = \dfrac{(-1)^{n+1} x^{n+1}}{(n)(4)^n}$, we have

$$\lim_{n\to\infty}\left|\frac{u_{n+1}}{u_n}\right| = \lim_{n\to\infty}\left|\frac{(-1)^{n+2} x^{n+2}}{(n+2)4^{n+1}} \cdot \frac{(n+1)(4)^n}{(-1)^{n+1} x^{n+1}}\right| =$$

$$\lim_{n\to\infty}\left|\frac{x}{4} \cdot \frac{n+1}{n+2}\right| = \left|\frac{x}{4}\right| \lim_{n\to\infty}\left(\frac{n+1}{n+2}\right) = \left|\frac{x}{4}\right|$$

$|\frac{x}{4}| < 1$

$|x| < 4$

$-4 < x < 4$

We now test at the endpoints.

When $x = 4$, we have $C + \frac{4}{2} - \frac{4}{3} + \frac{4}{4} - \frac{4}{5} + \ldots =$

$C + 4\left[\frac{1}{2} - \frac{1}{3} + \frac{1}{4} - \frac{1}{5} + \ldots\right]$ which is a convergent alternating series.

When $x = -4$, we have $C + \frac{4}{2} + \frac{4}{3} + \frac{4}{4} + \frac{4}{5} + \ldots =$

$C + 4\left[\frac{1}{2} + \frac{1}{3} + \frac{1}{4} + \frac{1}{5} + \ldots\right]$ which is a divergent harmonic series.

Therefore, the interval of convergence for $\int f(x)\, dx$ is also $(-4, 4]$.

Two series may be added, subtracted, multiplied, or divided (except by 0) for all x that are in the common interval of convergence for the two series.

Taylor Series

Let f be the function defined by the power series

$$f(x) = a_0 + a_1(x-c) + a_2(x-c)^2 + \ldots + a_n(x-c)^n + \ldots .$$

Then, the coefficients are given by

$$a_n = \frac{f^{(n)}(c)}{n!}$$

so

$$f(x) = f(c) + f'(c)(x-c) + \frac{f''(c)(x-c)^2}{2!} + \frac{f'''(c)(x-c)^3}{3!} + \ldots$$

$$+ \frac{f^n(c)(x-c)^n}{n!} + \ldots = \sum_{i=0}^{\infty} \frac{f^{(n)}(c)(x-c)^n}{n!}.$$

The above series is called a **Taylor Series** for $f(x)$ at $x = c$. If $c = 0$, the series is called a **Maclaurin Series**.

SEQUENCES AND SERIES

Functions can also be approximated by **Taylor polynomials**, $P_n(x)$, of order n such that

$$f(x) \approx P_n(x) = f(c) + f'(c)(x-c) + \frac{f''(c)(x-c)^2}{2!} + \ldots + \frac{f^n(c)(x-c)^n}{n!}$$

On the interval of convergence, the graphs of these Taylor polynomials become closer and closer approximations to the graph of f, as $n \to \infty$.

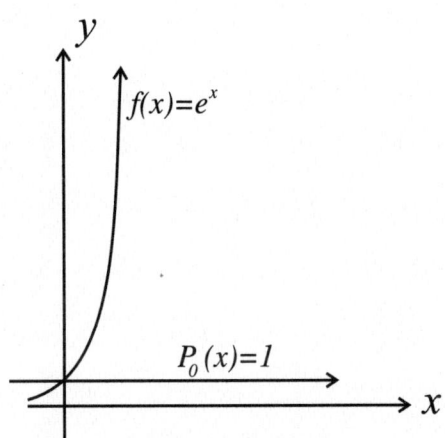

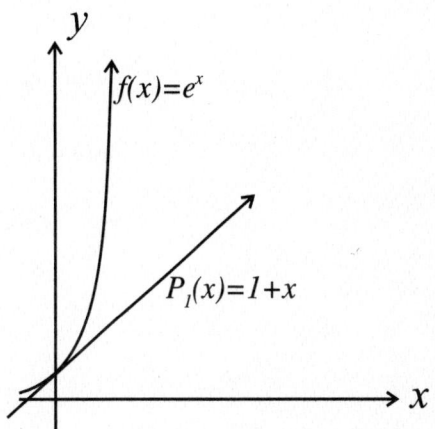

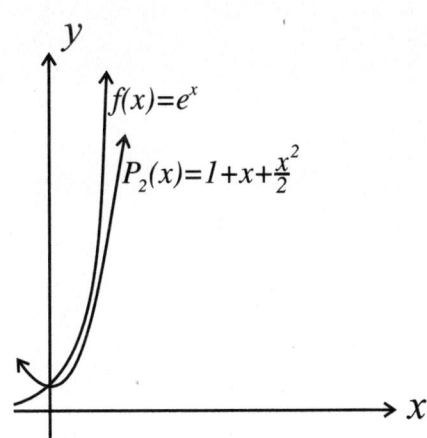

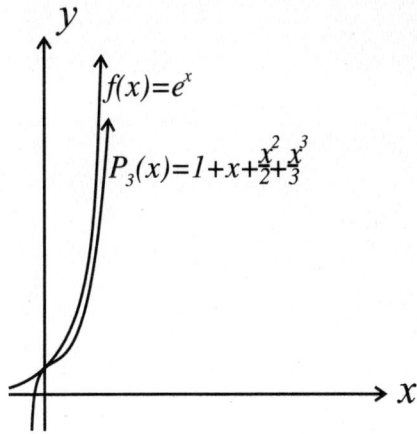

Example 40

Determine the Maclaurin polynomial of order 5 for $f(x) = \sin(x)$.

Solution:

$$\begin{aligned} f(x) &= \sin(x) & f(0) &= 0 \\ f'(x) &= \cos(x) & f'(0) &= 1 \\ f''(x) &= -\sin(x) & f''(0) &= 0 \\ f'''(x) &= -\cos(x) & f'''(0) &= -1 \\ f^{(iv)}(x) &= \sin(x) & f^{(iv)}(0) &= 0 \\ f^{(v)}(x) &= \cos(x) & f^{(v)}(0) &= 1 \end{aligned}$$

Therefore,

$$\sin(x) \approx P_5(x) = f(0) + f'(0)x + \frac{f''(0)x^2}{2!} + \frac{f'''(0)x^3}{3!} +$$

$$\frac{f^{(iv)}(0)x^4}{4!} + \frac{f^{(v)}(0)x^5}{5!}$$

$$= 0 + (1)x + \frac{0}{2!}x^2 - \frac{1}{3!}x^3 + \frac{0}{4!}x^4 + \frac{1}{5!}x^5$$

$$= x - \frac{x^3}{3!} + \frac{x^5}{5!}.$$

Example 41

Approximate $\sin(\pi/6)$ using $P_5(x)$ for $\sin(x)$.

Solution:

From Example 40, we have

$$\sin(x) \approx P_5(x) = x - \frac{x^3}{3!} + \frac{x^5}{5!}$$

Therefore, $P_n(\pi/6) = (\pi/6) - \dfrac{(\pi/6)^3}{3!} + \dfrac{(\pi/6)^5}{5!}$

$$\approx 0.52359 - 0.02392 + 0.00033$$

$$\approx 0.5$$

Hence, $P_5(x) \approx 0.5$.

SEQUENCES AND SERIES

Example 42

Determine the Maclaurin polynomial of order 4 for $f(x) = e^{-x}$.

Solution:

$$f(x) = e^{-x} \quad f(0) = 1$$
$$f'(x) = -e^{-x} \quad f'(0) = -1$$
$$f''(x) = e^{-x} \quad f''(0) = 1$$
$$f'''(x) = -e^{-x} \quad f'''(0) = -1$$
$$f^{(iv)}(x) = e^{-x} \quad f^{(iv)}(0) = 1$$

Therefore, $e^{-x} \approx$

$$P_4(x) = f(0) + f'(0)x + \frac{f''(0)x^2}{2!} + \frac{f'''(0)x^3}{3!} + \frac{f^{(iv)}(0)x^4}{4!}$$

$$= 1 + (-1)x + \frac{x^2}{2!} - \frac{x^3}{3!} + \frac{x^4}{4!}$$

$$= 1 - x + \frac{x^2}{2!} - \frac{x^3}{3!} + \frac{x^4}{4!}$$

Example 43

Determine the Maclaurin series for $f(x) = \cos(x)$.

Solution:

$$f(x) = \cos(x) \quad f(0) = 1$$
$$f'(x) = -\sin(x) \quad f'(0) = 0$$
$$f''(x) = -\cos(x) \quad f''(0) = -1$$
$$f'''(x) = \sin(x) \quad f'''(0) = 0$$
$$f^{(iv)}(x) = \cos(x) \quad f^{(iv)}(0) = 1$$
$$\vdots \quad \vdots$$

Hence,

$$\cos(x) = 1 + (0)x - \frac{x^2}{2!} + (0)x^3 + \frac{x^4}{4!} + \ldots + \frac{(-1)^n x^{2n}}{(2n)!} + \ldots$$

$$= 1 - \frac{x^2}{2!} + \frac{x^4}{4!} - \ldots + \frac{(-1)^n x^{2n}}{(2n)!} + \ldots$$

and can be shown to converge for all real values of x.

Example 44

Determine the Maclaurin series for $f(x) = \ln(x)$.

Solution:

$f(x) = \ln(x)$ and $f(0) = \ln(0)$ which does not exist. Therefore, there is no Maclaurin series for $\ln(x)$.

Example 45

Determine the Taylor series for $f(x) = \ln(x)$ at $x = 1$.

Solution:

$$f(x) = \ln(x) \qquad f(1) = \ln(1) = 0$$
$$f'(x) = \frac{1}{x} \qquad f'(1) = 1$$
$$f''(x) = \frac{-1}{x^2} \qquad f''(1) = -1$$
$$f'''(x) = \frac{2!}{x^3} \qquad f'''(1) = 2!$$
$$f^{(iv)}(x) = \frac{-3!}{x^4} \qquad f^{(iv)}(1) = -3!$$
$$\vdots \qquad \vdots$$
$$f^{(n)}(x) = \frac{(1)^{n+1}(n-1)!}{x^n} \qquad f^{(n)}(1) = (-1)^{n+1}(n-1)!$$
$$\vdots \qquad \vdots$$

Hence, $\ln(x) = (x-1) - \frac{(x-1)^2}{2} + \frac{(x-1)^3}{3} + \frac{(x-1)^4}{4} + \frac{(x-1)^5}{5} - \ldots$
$$+ \frac{(-1)^{n-1}(x-1)^n}{n} + \ldots$$

The interval of convergence for this series can be determined to be (0, 2).

SEQUENCES AND SERIES

The following are Maclaurin series for some elementary functions, together with their interval of convergence.

1. $\sin(x) = x - \dfrac{x^3}{3!} + \dfrac{x^5}{5!} - \dfrac{x^7}{7!} + \ldots + \dfrac{(-1)^n x^{2n+1}}{(2n+1)!} + \ldots \quad (-\infty, +\infty)$

2. $\cos(x) = 1 - \dfrac{x^2}{2!} + \dfrac{x^4}{4!} - \dfrac{x^6}{6!} + \ldots + \dfrac{(-1)^n x^{2n}}{(2n)!} + \ldots \quad (-\infty, +\infty)$

3. $e^x = 1 + \dfrac{x^2}{2!} + \dfrac{x^3}{3!} + \dfrac{x^4}{4!} + \ldots + \dfrac{x^n}{n!} + \ldots \quad (-\infty, +\infty)$

4. $\tan^{-1}(x) = x - \dfrac{x^3}{3} + \dfrac{x^5}{5} - \dfrac{x^7}{7} + \ldots + \dfrac{(-1)^n x^{2n+1}}{2n+1} + \ldots \quad [-1, 1]$

The following are Taylor series about $x = 1$ for some elementary functions, together with their interval of convergence.

1. $\ln(x) = (x-1) - \dfrac{(x-1)^2}{2} + \dfrac{(x-1)^3}{3} - \dfrac{(x-1)^4}{4} + \ldots + \dfrac{(-1)^{n-1}(x-1)^n}{n} + \ldots$
   (0, 2]

2. $\dfrac{1}{x} = 1 - (x-1) + (x-1)^2 - (x-1)^3 + (x-1)^4 - \ldots + (-1)^n (x-1)^n + \ldots$
   (0, 2)

3. $\dfrac{1}{x+1} = 1 - x + x^2 - x^3 + x^4 - x^5 + \ldots + (-1)^n x^n + \ldots$
   (−1, 1)

Example 46

Determine a power series for e^{x+2}.

Solution:

Using the Maclaurin series for e^x, we have

$$e^x = 1 + x + \dfrac{x^2}{2!} + \dfrac{x^3}{3!} + \dfrac{x^4}{4!} + \ldots + \dfrac{x^n}{n!} + \ldots \quad \text{for all } x \text{ in the}$$

interval $(\infty, +\infty)$. Hence,

$$e^{x+2} = 1 + (x+2) + \dfrac{(x+2)^2}{2!} + \dfrac{(x+2)^3}{3!} + \dfrac{(x+2)^4}{4!} + \ldots$$

$$+ \dfrac{(x+2)^n}{n!} + \ldots \ . \text{ Its interval of convergence is } (-\infty, +\infty).$$

Example 47
Determine a Power series for e^{x+2}.

Solution (Alternate):
We first consider $e^{x+2} = (e^x)(e^2)$. Using the Maclaurin series for e^x, we have

$$e^x = 1 + x + \frac{x^2}{2!} + \frac{x^3}{3!} + \frac{x^4}{4!} + \ldots + \frac{x^n}{n!} + \ldots .$$

Next, multiply by e^2, obtaining

$$e^{x+2} = e^2 \cdot e^x = e^2 \sum_{n=0}^{\infty} \frac{x^n}{n!} .$$

Example 48
Starting with the Maclaurin series for $\sin(x)$, determine the Maclaurin series for $\cos(x)$.

Solution:

$$\sin(x) = x - \frac{x^3}{3!} + \frac{x^5}{5!} - \frac{x^7}{7!} + \ldots + \frac{(-1)^n x^{2n+1}}{(2n+1)!} + \ldots .$$

Since $\cos(x) = \frac{d}{dx}(\sin(x))$, we have

$$\cos(x) = \frac{d}{dx}\left(x - \frac{x^3}{3!} + \frac{x^5}{5!} - \frac{x^7}{7!} + \ldots + \frac{(-1)^n x^{2n+1}}{(2n+1)!} + \ldots \right)$$

$$= 1 - \frac{x^2}{2!} + \frac{x^4}{4!} - \frac{x^6}{6!} + \ldots + \frac{(-1)^n x^{2n}}{(2n)!} + \ldots .$$

which is the Maclaurin series for $\cos(x)$.

SEQUENCES AND SERIES

LAGRANGE FORM OF REMAINDER IN TAYLOR SERIES

Let f be a function with $n + 1$ derivatives in some interval containing c. If x is any number in the interval $x \neq c$, then the error made in approximating $f(x)$ by

$$P_n(x) = f(c) + f'(c)(x-c) + \frac{f''(c)(x-c)^2}{2!} + \ldots + \frac{f^{(n)}(c)(x-c)^n}{n!}$$

is less than $|R_n(x)|$ where

$$R_n(x) = \frac{f^{(n+1)}(t)(x-c)^{n+1}}{(n+1)!}$$

for some t between c and x.

Example 49

Let $f(x) = \ln(x)$. Approximate $\ln(1.1)$ to four decimal places by using $P_3(1.1)$ and $R_3(1.1)$.

Solution:
Using the Taylor series for $\ln(x)$ about $x = 1$, we have

$\ln(x) \approx$

$$P_3(x) = f(1) + f'(1)(x-1) + \frac{f''(1)(x-1)^2}{2!} + \frac{f'''(1)(x-1)^3}{3!}$$

$$= 0 + (1)(x-1) + \frac{(-1)(x-1)^2}{2!} + \frac{2(x-1)^3}{3!}$$

Hence,

$$\ln(1.1) \approx P_3(1.1) = (0.1) - \frac{(0.1)^2}{2} + \frac{(0.1)^3}{3}$$

or, $\ln(1.1) \approx 0.0953$.

$$R_3(x) = \frac{(-6)(x-1)^4}{4! \; t^4}$$

where t is between 1 and x.

Therefore,

$$|R_3(1.1)| = \left| \frac{-(0.1)^4}{4t^4} \right| \text{ where } 1 < t < 1.1 \;.$$

Since $t > 1$, then $\frac{1}{t} < 1$ and $\frac{1}{t^4} < 1$. Therefore,

PART IV

$$|R_3(1.1)| = \left|\frac{-(0.1)^4}{4t^4}\right| < \left|\frac{-0.0001}{4}\right| = 0.000025 < 0.00005.$$

Therefore, $\ln(1.1) \approx 0.0953$ is accurate to four decimal places.

Example 50

Compute $\sin(34°)$ correct to four decimal places.

Solution:

Using the Taylor series for $\sin(x)$ about $x = \pi/6$, we have $\sin(x) \approx$

$$P_3(x) = \frac{1}{2} + \frac{\sqrt{3}}{2}(x - \pi/6) - \frac{1}{(2)(2!)}(x - \pi/6)^2 - \frac{\sqrt{3}}{(2)(3!)}(x - \pi/6)^3$$

$$= \frac{1}{2} + \frac{\sqrt{3}}{2}(x - \pi/6) - \frac{1}{4}(x - \pi/6)^2 - \frac{\sqrt{3}}{12}(x - \pi/6)^3$$

Since $x = 34°$, $x - \pi/6 = 34° - 30° = 4°$ and then, converting to radians, $4° = 4(0.01745) = 0.0698$.

Hence, $\sin(34°) \approx$

$$P_3(0.0698) = \frac{1}{2} + \frac{\sqrt{3}}{2}(0.0698) - \frac{1}{4}(0.0698)^2 - \frac{\sqrt{3}}{12}(0.0698)^3$$

$$\approx 0.5 + 0.0604 - 0.00122 - 0.00005$$

$$\approx 0.5592.$$

$$R_3(x) = \frac{f^{(iv)}(t)}{4!}(x - \pi/6)^4$$

$$= \frac{\sin(t)}{4!}(x - \pi/6)^4$$

where $\pi/6 < t < 0.5934$ ($0.5934 = 34\pi/180$)

and $|R_n(34°)| = \left|\frac{\sin(t)}{4!}(0.0698)^4\right|$

$$\leq \left|\frac{\sqrt{2}}{(2)(4!)}(0.0698)^4\right|$$

$$\leq \left|\frac{\sqrt{2}}{48}(0.0698)^4\right|$$

$$= 0.00000069935.$$

Therefore, $\sin(34°) \approx 0.5592$ correct to four decimal places.

ADVANCED PLACEMENT CALCULUS AB

PRACTICE TEST 1

Time	Number of Questions	Percent of Total Grade
1 hour and 30 minutes	40	50

SECTION I, PART A

Directions: Solve each of the following problems using the available space for scratchwork. After examining the form of the choices, decide which is the best of the choices given and fill in the corresponding oval on the answer sheet. No credit will be given for anything written in the test book. Do not spend too much time on any one problem.

<u>In this test</u>: Unless otherwise specified, the domain of a function f is assumed to be the set of all real numbers x for which $f(x)$ is a real number.

1. If $f(x) = 6x^{3/2}$, then $f''(16) =$

   (A) 9/16
   (B) 9/8
   (C) 36
   (D) 144
   (E) 384

2. The slope of $9x - 4x \ln y = 3$ at $(1/3, 1)$ is

   (A) $9 - 4 \ln 3$
   (B) 5
   (C) 6
   (D) 27/4
   (E) $9 + 4 \ln 3$

3. If $g(x) = 8x^3 - 3x^2 + 4$, then a relative minimum occurs at

   (A) $(-1/2, 9/4)$
   (B) $(0, 4)$
   (C) $(1/8, 127/32)$
   (D) $(1/4, 63/16)$
   (E) $(0, 4)$ and $(1/4, 63/16)$

4. $\int x\sqrt{3x^2 + 4}\, dx$

   (A) $(1/9)(3x^2 + 4)^{3/2} + C$
   (B) $(2/3)(3x^2 + 4)^{3/2} + C$
   (C) $2(3x^2 + 4)^{1/2} + C$
   (D) $6x(3x^2 + 4) + C$
   (E) $(3/5)x^5 + (4/3)x^3 + C$

5. The equation $y = 6 \sin(3x - 1)\pi + 4$ has a fundamental period of

   (A) $1/3$
   (B) $2/3$
   (C) $2\pi/3$
   (D) 4
   (E) 6

6. $\lim_{x \to \infty} \dfrac{3x^2 - 5x + 4}{6x^2 + 7x - 1}$

   (A) -4
   (B) $-5/7$
   (C) 0
   (D) $1/6$
   (E) $1/2$

7. If the function f is continuous for all real numbers and if $f(x) = (2x^2 + x - 15)/(x + 3)$ when $x \neq -3$, then $f(-3) =$

   (A) -15
   (B) -11
   (C) -3
   (D) 0
   (E) $5/2$

8. If $h(x) = (6x - 3x^2)/(4 + x)$, then $h'(x) =$

   (A) $6 - 6x$
   (B) $3/2 - 3x$
   (C) -3
   (D) $(-3x^2 - 24x - 24)/(4 + x)^2$
   (E) $[-3(x^2 + 8x - 8)]/(x + 4)^2$

9. $\int_{-1}^{1} \frac{6}{1+x^2} dx$

   (A) 0
   (B) 3
   (C) π
   (D) 2π
   (E) 3π

10. If $g(x) = \csc x - \cot x$, then $g'(\pi/6) =$

    (A) $4 - 2\sqrt{3}$
    (B) $2 - 2\sqrt{3}$
    (C) $2 - \sqrt{3}$
    (D) 1
    (E) $4 + \sqrt{3}$

11. If $s(x) = \sin^2 x$, then $s''(x) =$

    (A) -2
    (B) $-2 \cos x \sin x$
    (C) $2 \sin x \cos x$
    (D) $2 \cos^2 x - 2 \sin^2 x$
    (E) 2

12. If $f(x) = 5/(x^3 - 1)$ and $g(x) = 2x$, then $f(g(-1)) =$

    (A) $-5/2$
    (B) -2
    (C) $-5/9$
    (D) $5/6$
    (E) undefined

13. The average value of $(3x + 1)^2$ on the interval -1 to 1 is

    (A) $-1/9$
    (B) $1/9$
    (C) $2/9$
    (D) 4
    (E) 8

14. If $g(x) = 2^{4x}$, then $g''(x) =$

    (A) 8
    (B) $4(\ln 2)$
    (C) $4 * \ln 2 * 2^{4x}$
    (D) $16^{x+1}(\ln 2)^2$
    (E) $16^x(\ln 2)^2$

15. $\lim\limits_{x \to 0} \dfrac{-\cos\ x + 1 - \sin\ x}{x}$

    (A) -1
    (B) 0
    (C) undefined
    (D) 1
    (E) ∞

16. $y' = \dfrac{4x}{y}$ and $x = -1$ when $y = 4$. What can x be when $y = 6$?

    (A) -6
    (B) $-\sqrt{6}$
    (C) -2
    (D) 2
    (E) 6

17. Given $f(x) = \sin x + \cos x$ on the interval $[0,\pi]$, what is the absolute maximum function value?

    (A) $-\sqrt{2}$
    (B) -1
    (C) 0
    (D) 1
    (E) $\sqrt{2}$

18. $\int e^x x^2 \, dx$

   (A) $x^2 e^x + 2x e^x + 2e^x + C$
   (B) $2x e^x$
   (C) $2x e^x + C$
   (D) $2x e^x + e^x x^2 + C$
   (E) $e^x(x^2 - 2x + 2) + C$

19. The Panda Bear Club realizes a monthly revenue of $R(x) = 5{,}250x - 10x^2$ dollars per month when the fee per person is x dollars. What is the marginal revenue when the fee is $21 per person?

   (A) –$20
   (B) $4,830
   (C) $61,560
   (D) $105,840
   (E) $1,126,755

20. What is the slope of the tangent line to the curve $y = \cos^2(3x)$ at the point $(\pi/4, 1/2)$?

   (A) –3
   (B) 1/2
   (C) 0
   (D) 2
   (E) 3

21. If $y = \ln(x^2 - 5)$, what is $y''(4)$?

   (A) –42/121
   (B) 8/11
   (C) 2
   (D) 2.398
   (E) 8

22. Where is the function $f(x) = 5/(x^2 - 2x - 15)$ discontinuous?

   (A) $x = -5$ and $x = -3$
   (B) $x = -5$ and $x = 3$
   (C) $x = -3$ and $x = 5$
   (D) $x = 3$ and $x = 5$
   (E) $x = 2$ and $x = 15$

23. ln 2 + ln 5 − ln 8 − ln 15 =

   (A) − ln 12
   (B) ln 12
   (C) ln 7 − ln 23
   (D) − ln (1/12)
   (E) ln (−16)

24. $-\sin^2 x + 4 - \cos^2 x =$

   (A) −5
   (B) −3
   (C) 3
   (D) 5
   (E) cannot be determined

25. Given $y = \sqrt{x^3}$, what is $y'''(4)$?

   (A) −4
   (B) −3/64
   (C) −1/48
   (D) 1/6
   (E) 4

Section I, Part B

A GRAPHING CALCULATOR IS REQUIRED FOR SOME QUESTIONS ON THIS PART OF THE EXAMINATION.

Directions: Solve each of the following problems using the available space for scratchwork. After examining the form of the choices, decide which is the best of the choices given and fill in the corresponding oval on the answer sheet. No credit will be given for anything written in the test book. Do not spend too much time on any one problem.

In this test:

(1) The *exact* numerical value of the correct answer does not always appear among the choices given. When this happens, select from among the choices the number that best approximates the exact numerical value.

(2) Unless otherwise specified, the domain of a function f is assumed to be the set of all real numbers x for which $f(x)$ is a real number.

26. $\int_{\pi}^{5\pi/4} \tan x \, dx + \int_{-\pi}^{\pi} \cos x \, dx$

   (A) $-\ln \sqrt{2}$
   (B) $\ln 2 + \ln \sqrt{2}$
   (C) 1
   (D) $\ln 2 - \ln \sqrt{2}$
   (E) undefined

27. $\lim\limits_{h \to 0} \dfrac{\sec(\pi/3 + h) - \sec(\pi/3)}{h}$

   (A) 1/2
   (B) $\sqrt{3}$
   (C) 2
   (D) $2\sqrt{3}$
   (E) undefined

ADVANCED PLACEMENT CALCULUS AB

28. What is the volume of the solid obtained when the region bounded by $x = 4$, $x = 9$, $y=0$, and $y = \sqrt{x}$ is rotated about the x-axis?

   (A) 1
   (B) 2
   (C) $65\pi/2$
   (D) 65
   (E) 65π

29. $\dfrac{d}{dx}\left(\displaystyle\int_0^{x/2} \sin(t)dt\right)$

   (A) $1/4 \sin(x/2)$
   (B) $1/2 \sin(x/2)$
   (C) $\sin(x/2)$
   (D) $2 \sin(x/2)$
   (E) 0

30. A showroom is to be made in the shape of a rectangle. The perimeter of the showroom must be 1,200 feet. What is the maximum area the showroom can have?

   (A) 22,500 square feet
   (B) 25,000 square feet
   (C) 90,000 square feet
   (D) 200,000 square feet
   (E) 250,000 square feet

31. $\displaystyle\int \dfrac{x\,dx}{\sqrt{(3+x)}}$

   (A) $(x^2/2)(3 + x)^{-3/2} + C$
   (B) $2x^2(3 + x)^{-3/2} + C$
   (C) $2x^2(3 + x)^{3/2} + C$
   (D) $[2(x - 6)(3 + x)^{1/2}]/3 + C$
   (E) $\left[2(x - 8)\sqrt{\dfrac{(1+x)}{3}}\right] + C$

32. What is x when $6 = e^{5x}$?

   (A) $e^6/5$
   (B) $6 - \ln 5$
   (C) $5 \ln 6$
   (D) $(\ln 6)/5$
   (E) $6/(e^5)$

33. What is the equation of the line that is tangent to the curve $M(t) = -t^2 + 2t + 4$ at the point $(-1, 1)$?

   (A) $M = t + 1$
   (B) $M = 3t/5 + 52/5$
   (C) $M = t + 2$
   (D) $M = 3t/5 - 2$
   (E) $M = 4t + 5$

34. What is the equation of the normal line to $y = \dfrac{x+5}{x-5}$ at $y = 11$?

   (A) $y = x/10 - 52/5$
   (B) $y = x/10 + 5$
   (C) $x + 10y = 104$
   (D) $15y - 54x + 554 = 0$
   (E) $x - 10y = -104$

35. The diameter of a circle is increasing at a rate of 5 cm/sec. At what rate is the area increasing when the diameter is 10 cm?

   (A) 6.25π cm²/sec.
   (B) 25π cm²/sec.
   (C) 50π cm²/sec.
   (D) $125\pi/3$ cm²/sec.
   (E) 100π cm²/sec.

36. If $y = (x + 5)(3x)$, what is y'?

   (A) $3x + 15$
   (B) 3
   (C) 4
   (D) $6x + 15$
   (E) $3 + 3x$

37. Where does the graph of $f(x) = 3x^2 - 24x + 36$ change direction?

   (A) $x = 2$ and $x = 6$
   (B) $x = 2$
   (C) $x = 6$
   (D) $x = 4$ and $x = 6$
   (E) $x = 4$

38. Given the parametric equations $x = t$ and $y = \sqrt{1-t^2}$, what is the Cartesian equation?

   (A) $y = 1 - x^2$
   (B) $y = \sqrt{(1-x)}$
   (C) $x^2 + y^2 = 1$
   (D) $x + y = 1$
   (E) $y = x$

39. $\lim\limits_{x \to 0} \dfrac{2\sin x \cos x}{2x}$

   (A) -2
   (B) -1
   (C) undefined
   (D) $1/2$
   (E) 1

40. $\lim\limits_{x \to 2} \dfrac{4x^2 - 16}{(x-2)} =$

   (A) -3
   (B) $-1/4$
   (C) -1
   (D) 0
   (E) 16

END OF SECTION I

IF YOU FINISH BEFORE TIME IS CALLED, YOU MAY CHECK YOUR WORK ON THIS SECTION.
DO NOT GO TO SECTION II UNTIL YOU ARE TOLD TO DO SO.

Section II

SHOW ALL YOUR WORK. Indicate clearly the methods you use, because you will be graded on the correctness of your methods as well as on the accuracy of your final answers. If you choose to use decimal approximations, your answer should be correct to three decimal places.

A GRAPHING CALCULATOR IS REQUIRED FOR SOME QUESTIONS ON THIS PART OF THE EXAMINATION.

Note: Unless otherwise specified, the domain of a function f is assumed to be the set of all real numbers x for which $f(x)$ is a real number.

1. $f(x) = x^2$ and $g(x) = \sqrt{x}$?

   (A) What are the points of intersection of the two curves?
   (B) What is the area of the region bounded by the two curves?
   (C) What is the volume of the region bounded by the two curves rotated about the x-axis?
   (D) What is the volume of the region bounded by the two curves rotated about the y-axis?

2. A family standing on the ground is watching a jet approach. The jet is approaching at a rate of 12 miles per minute at an altitude of 8 miles.

   (A) At what speed, (in radians per minute), is the angle of altitude changing when the straight-line distance from the family to the jet is 36 miles?
   (B) At what rate is the straight-line distance between the family and the jet changing when the jet is 16 miles from the family?

3. Suppose $x^3 - 2x^2y + y^2 = -1$.

   (A) What is the equation of the line tangent to this curve at $(-1, 2)$?
   (B) What is the equation of the normal line at $y = 2$, $x = -1$?
   (C) When is the tangent line vertical?

4. Let f be the function given by $f(x) = 4x^3 + 6x^2 - 5$

   (A) Identify any relative maxima or minima.
   (B) Identify any inflection point(s).
   (C) What is the equation of the line tangent to the curve at the point (-2,-13)?
   (D) Sketch the graph.

5. A 10-foot ladder leans against a building. A boy pulls the base of the ladder out from the wall at the rate of 2 feet per second.

   (A) At what rate is the ladder moving down the building when the foot of the ladder is 5 feet from the building?
   (B) At what rate does the angle between the ladder and the ground change when that angle is $\pi/4$ radians?

6. A bacteria culture grows from 100 to 300 grams in 6 hours following a natural, exponential growth.
   (A) How much bacteria is present after five hours?
   (B) How long did it take to double the bacteria population?
   (C) When will there be 1,000 grams?

ADVANCED PLACEMENT CALCULUS AB

PRACTICE TEST 2

Time	Number of Questions	Percent of Total Grade
1 hour and 30 minutes	40	50

SECTION I, PART A

Directions: Solve each of the following problems using the available space for scratchwork. After examining the form of the choices, decide which is the best of the choices given and fill in the corresponding oval on the answer sheet. No credit will be given for anything written in the test book. Do not spend too much time on any one problem.

<u>In this test</u>: Unless otherwise specified, the domain of a function f is assumed to be the set of all real numbers x for which $f(x)$ is a real number.

---

1. If $f(x) = 6x^2 - 4x + 2$, then $f^{-1}(4) =$

    (A) 1/82
    (B) 1/78
    (C) 1
    (D) 11/8
    (E) 2

---

2. $h(x) = \begin{cases} x^2 - 1 & \text{if } x < 0 \\ x - 1 & \text{if } 0 \leq x \leq 3 \\ x^3 & \text{if } x > 3 \end{cases}$

    Where is h discontinuous?

    (A) $x = 0$
    (B) $x = 0$ and $x = 3$
    (C) $x = \pm 1$
    (D) $x = 3$
    (E) $x = \pm 1$ and $x = 3$

ADVANCED PLACEMENT CALCULUS AB

3. $g(x) = \dfrac{(x+5)}{(x^2 - 25)}$

   What is the domain?

   (A) all real numbers
   (B) all real numbers, $x \neq \pm 5$
   (C) all real numbers, $x \neq 5$
   (D) all real numbers, $x \neq -5$
   (E) all real numbers, $x \neq 25$

4. $f(x) = x^3 + 2x \qquad g(x) = \sqrt{x-1}$

   What is $g(f(2))$?

   (A) -1
   (B) $\sqrt{11}$
   (C) 1
   (D) 3
   (E) 12

5. The Graham Company's cost to produce a rug is $250. The company's fixed cost per month is $12,000. They sell the rugs for $350 each. In one month's time, how many rugs do they need to produce to break even?

   (A) 1.4 rugs
   (B) 34.29 rugs
   (C) 48 rugs
   (D) 120 rugs
   (E) 1,200 rugs

6. Given the function $h(x) = 6x^3 - 8x^2 + 2$, at what x value(s) is/are the inflection point(s)?

   (A) $x = 4/9$
   (B) $x = 0$ and $x = 8/9$
   (C) $x = 0$
   (D) $x = 0$, $x = 4/9$, and $x = 8/9$
   (E) $x = 0$ and $x = 4/9$

7. What is the average value of $f(x) = e^{2x}$ over the interval $[1,4]$?

   (A) $e^8 - e^2$
   (B) e^6
   (C) e^4
   (D) $(e^8 - e^2)/6$
   (E) $2(e^8 - e^2)/3$

8. Given $yy' = \dfrac{1}{x}$ and $y(2) = 1$, then $y^2 =$

   (A) x^{-2}
   (B) 4
   (C) $4 + 2\ln|x|$
   (D) $2\ln|x| + 1 - \ln 4$
   (E) $2\ln|x|$

9. Hal invests $1000 in an account that earns interest compounded continuously. The annual interest rate is 5.25%. How much does Hal have in his account after six years?

   (A) $1,020.94
   (B) $1,052.50
   (C) $1,315
   (D) $1,359.35
   (E) $1,370.26

10. Given $w = (3z^2 - 5z)^5$, what is w'?

    (A) $15z - 5$
    (B) $(30z - 5)(3z^2 - 5z)^4(6z - 5)$
    (C) $5(6z - 5)(3z^2 - 5z)^4$
    (D) $250z^9 - 3125z^4$
    (E) $5(3z^2 - 5z)$

11. Water flows into a conical tank at a constant rate of 3 cubic meters per second. The radius of the cone is 5 meters and its height is 6 meters. At what rate is the water level rising when the radius is 4 meters?

    (A) $3/(16\pi)$ meters per second
    (B) $9/(100\pi)$ meters per second
    (C) .8 meters per second
    (D) $108/(25\pi)$ meters per second
    (E) 3 meters per second

12. If $y' = \dfrac{x^3}{y^2}$, then $y^3 =$

   (A) $x^4 + C$
   (B) $12x^4 + C$
   (C) $\dfrac{3}{4}x^4 + C$
   (D) $\dfrac{4}{3}x^4 + C$
   (E) $x^3y + 12C$

13. The line through the point (–5,4) and perpendicular to $4x - 3y = 5$ has y-intercept

   (A) –2
   (B) –5/3
   (C) –1/4
   (D) 1/4
   (E) 5

14. The graph of $y^2 + 3x - 2y + 8 = 7 - x^2$ is

   (A) an ellipse
   (B) a pair of straight lines
   (C) a parabola
   (D) hyperbola
   (E) a circle

15. If $f(u) = \ln u$ and $g(u) = e^{3u}$, the $g(f(1)) =$

   (A) undefined
   (B) 0
   (C) 1
   (D) 3
   (E) e^{3u}

16. $\lim\limits_{x \to 3} \dfrac{6x^2 - 5}{4x^2 + 1}$

    (A) –5

    (B) $\dfrac{1}{5}$

    (C) $\dfrac{49}{37}$

    (D) $\dfrac{3}{2}$

    (E) $\dfrac{13}{5}$

17. If the position of a particle at any time t is given by $s = -t^3 - 5t$, then the speed of the particle at time $t = 7$ is

    (A) –378
    (B) –152
    (C) –42
    (D) 42
    (E) 152

18. The average value of $y = (2x + 5)^3$ over the interval (1,4) is

    (A) 360
    (B) 618
    (C) 927
    (D) 1090
    (E) 6540

19. Given $f(x) = 1/(3x + 5)$, $f^{-1}(x) =$

    (A) $3x + 5$
    (B) $(1 - 5x)/3x$
    (C) $1/(3x - 5)$
    (D) $3x - 5$
    (E) $1/2$

20. The graph of $50y + 25y^2 = 96x - 16x^2 + 231$ is symmetric with

    (A) the x-axis
    (B) the y-axis
    (C) the origin
    (D) $y = x$
    (E) none of these

21. The area bounded by $y = x^3$, $y = 0$, $x = 2$, and $x = 7$ is

    (A) 67
    (B) 119.25
    (C) 335
    (D) 343
    (E) 596.25

22. How many zeros does $f(x) = 6x^4 - 8x^3 + 7x^2 - 5x + 9$ have?

    (A) 0
    (B) 1
    (C) 2
    (D) 3
    (E) 4

23. The curve of $y = (x - 1)/(3 + x)$ is concave down over

    (A) $(-3, \infty)$
    (B) $(-\infty, -3)$
    (C) $(-\infty, -3) \cup (-3, \infty)$
    (D) all real numbers
    (E) never

24. $\int_1^2 (5x - 3)^2 \, dx =$

    (A) 45
    (B) 67/3
    (C) 45
    (D) 335/3
    (E) 335

25. $\lim_{x \to 5} \dfrac{x^2 - x - 20}{x - 5}$

   (A) undefined
   (B) 1
   (C) 5
   (D) 9
   (E) 10

Section I, Part B

A GRAPHING CALCULATOR IS REQUIRED FOR SOME QUESTIONS ON THIS PART OF THE EXAMINATION.

Directions: Solve each of the following problems using the available space for scratchwork. After examining the form of the choices, decide which is the best of the choices given and fill in the corresponding oval on the answer sheet. No credit will be given for anything written in the test book. Do not spend too much time on any one problem.

In this test: The *exact* numerical value of the correct answer does not always appear among the choices given. When this happens, select the number that best approximates the exact numerical value.

Unless otherwise specified, the domain of a function f is assumed to be the set of all real numbers x for which $f(x)$ is a real number.

26. The equation of the normal line to the curve $y = (x - 5)/(x + 3)$ at the point $(-1,-3)$ is

   (A) $y = 1/2x - 5$
   (B) $x - 2y = -3$
   (C) $x + 2y = -7$
   (D) $y = 2x + 3$
   (E) $y = 2x - 6$

27. If $y = uv$ and u and v are both differentiable, then uy'' is

   (A) $u^2 v'' + uu''v$
   (B) $u^2 v'' + 2uu'v' + uu''v$
   (C) $2uv + u^2v'$
   (D) $uv' + u'v$
   (E) $u^2v' + uu'v$

28. Suppose that a ball thrown from the surface of the earth with an initial velocity of 128 ft/sec attains a height of $(-16t^2 + 128t)$ ft at time t. The ball's velocity in feet per second when it hits the ground is

   (A) -128 ft/sec
   (B) -96 ft/sec
   (C) -32 ft/sec
   (D) 96 ft/sec
   (E) 128 ft/sec

29. How many inflection points does $3x^4 - 5x^3 - 9x + 2$ have?

   (A) 0
   (B) 1
   (C) 2
   (D) 3
   (E) 4

30. $\int x^3(x - \sqrt{x} + 2)\ dx$

   (A) $4x^3 - (7/2)x^{5/2} + 6x^2 + C$
   (B) $x^4/4[x^2/2 - (2/3)x^{3/2} + 2x]$
   (C) $x^5/5 - (2/9)x^{9/2} + x^5/2$
   (D) $3x^2[1 - (1/2)x^{-1/2}] + C$
   (E) $x^5/5 - (2/9)x^{9/2} + x^4/2 + C$

31. If $f(\theta) = 6\cot(2\theta)$, then $f'(\theta) =$

   (A) $-6\csc^2(2\theta)$
   (B) $12\csc(2\theta)$
   (C) $-12\sin(2\theta)$
   (D) $6\csc^2(2\theta)$
   (E) $-12\csc^2(2\theta)$

32. $\lim\limits_{h \to 0} \dfrac{\cot\left(\dfrac{5\pi}{6} + h\right) - \cot\dfrac{5\pi}{6}}{h}$

   (A) -4
   (B) $-\sqrt{3}$
   (C) $\sqrt{3}$
   (D) 4
   (E) cannot be determined

33. $\int_0^{\pi/4} \sin x\, dx - \int_{-\pi}^{\pi} \cos x\, dx$

   (A) $(1 - \sqrt{2})/2$
   (B) $(2 - \sqrt{2})/2$
   (C) $(1 + \sqrt{2})/2$
   (D) $(2 + \sqrt{2})/2$
   (E) $2 + \sqrt{2}$

34. If $6y = 3e^{2x}$, then $y' =$

   (A) $(1/2)e^{2x}$
   (B) $3e^x$
   (C) $3e^{2x}$
   (D) $6e^x$
   (E) e^{2x}

35. $\text{Arccos}[\sin(7\pi/6)] =$

   (A) $-1/2$
   (B) $-\pi/3$
   (C) $\pi/3$
   (D) $2\pi/3$
   (E) $5\pi/6$

36. If $y = \tan(3x - 5\pi)$, then $y'(\pi) =$

   (A) -3
   (B) -2
   (C) 0
   (D) 2
   (E) 3

37. The circumference of a circle is changing at a rate of 10 cm/sec. At what rate is the radius of the circle changing when the diameter is 5 cm?

   (A) 2.5 cm/sec.
   (B) $5/\pi$ cm/sec.
   (C) $10/\pi$ cm/sec.
   (D) 5π cm/sec.
   (E) 10π cm/sec.

38. The function graphed below is NOT differentiable when

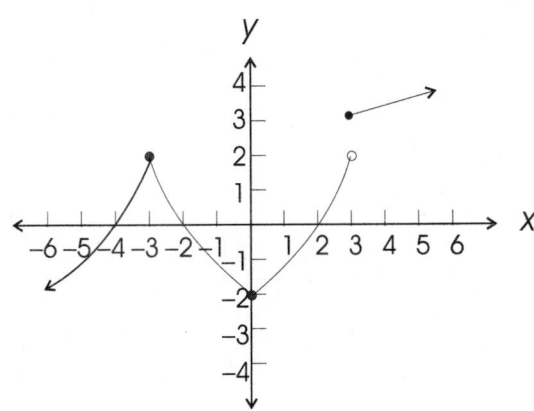

   (A) $x = 3$
   (B) $x = -6$ and $x = 3$
   (C) $x = -3$, $x = 3$, and $x = 0$
   (D) $x = -3$
   (E) $x = -6$, $x = -3$, $x = 0$, $x = 3$, and $x = 6$

39. If $f(x) = 5x^4 - 2x^3 + 3$ and $g(x) = f^{-1}(x)$, then $g'(10)$ is

   (A) -26
   (B) $-1/26$
   (C) 3
   (D) 19,400
   (E) 48,003

40. Given $y(x) = \ln e^{6 \csc x + 5x}$, $y'(x) =$

   (A) $(-6 \csc x \cot x + 5)(e^{6 \csc x + 5})$
   (B) $-6 \csc x \cot x + 5$
   (C) $-6 \csc x \cot x$
   (D) $(-6 \csc x \cot x + 5)/(e^{6 \csc x + 5x})$
   (E) $(6 + 5 \sin x)/(\sin x)$

END OF SECTION I

IF YOU FINISH BEFORE TIME IS CALLED, YOU MAY CHECK YOUR WORK ON THIS SECTION.
DO NOT GO TO SECTION II UNTIL YOU ARE TOLD TO DO SO.

PRACTICE TEST 2, SECTION II

Time	Number of Questions	Percent of Total Grade
1 hour and 30 minutes	6	50

SECTION II

SHOW ALL YOUR WORK. Indicate clearly the methods you use because you will be graded on the correctness of your methods as well as on the accuracy of your final answers. If you choose to use decimal approximations, your answer should be correct to three decimal places.

A GRAPHING CALCULATOR IS REQUIRED FOR SOME QUESTIONS
ON THIS PART OF THE EXAMINATION.

Note: Unless otherwise specified, the domain of a function f is assumed to be the set of all real numbers x for which $f(x)$ is a real number.

---

1. Sasha Connor was born today. She will start college in eighteen years.

   (A) How much money should her grandparents put into an account today to assure that she will have $75,000 in it when she starts her college career? The account earns 7.75% annual interest compounded continuously.

   (B) What interest rate should the grandparents find to end up with the same amount of money after only fifteen years, starting with the same original investment as in (A)?

   (C) How many years would the original invested amount have to stay in the account earning 7.75% interest if Sasha will need $45,000?

---

2. $f(x) = \cos x$ and $g(x) = \sin x$

   (A) What is the area bounded by the curves between $x = 0$ and $x = \pi/4$?
   (B) What is the volume of the solid obtained by rotating this region bounded by the curves about the x-axis?
   (C) What is the volume of the solid obtained by rotating this region bounded by the curves about the y-axis?
   (D) What is the volume of the solid obtained by rotating this region bounded by the curves about the line $x = -1$?

3. Find each limit:

   (A) $\lim_{x \to 5^+} (x-5)$

   (B) $\lim_{x \to \infty} \ln(1+x)$

   (C) $\lim_{x \to 0^+} (\sin x)$

4. Sketch the graph of a function that satisfies ALL of the following conditions:

   $f(0) = 0, f(-1) = 1, f'(-1) = 0, f''(x) > 0$ on $(-\infty, -1), f''(x) < 0$ on $(-1, 0)$ and $(0, \infty), f'(x) > 0$ for $x > 0$

5. Four hundred bacteria are present in a culture after two hours. After six hours, the culture contains 25,600 bacteria.

   (A) How many bacteria were in the culture at the beginning?
   (B) How long does it take this population to double?
   (C) When will this population reach 50,000?

6. A football team plays in a stadium that has a seating capacity of 15,000 spectators. If tickets cost $12, then attendance at the game is 11,000 spectators. An analysis indicates that for every dollar decrease in ticket price, the attendance at the game increases by 1,000 spectators. What ticket price should be set to maximize the revenue from ticket sales?

ADVANCED PLACEMENT CALCULUS BC

PRACTICE TEST 1

Time	Number of Questions
55 Minutes	28

SECTION I, PART A

Do not spend too much time on any one problem.

Directions: Solve each of the following problems using the available space for scratchwork. After examining the form of the choices, decide which is the best of the choices given and fill in the corresponding oval on the answer sheet. No credit will be given for anything written in the test book.

A CALCULATOR MAY NOT BE USED ON THIS PART OF THE EXAMINATION.

In this test: Unless otherwise specified, the domain of a function f is assumed to be the set of all real numbers x for which $f(x)$ is a real number.

---

1. Let $f(x)$ be a function such that $\int_1^5 f(x)dx = 2$. What is $\int_1^5 2f(x) - 1\,dx$?
    (A) 0
    (B) 1
    (C) 2
    (D) 3
    (E) 4

---

2. What is $\int_0^{\frac{\pi}{6}} \cos x(\sin^2 x + 1)dx$?

    (A) $\dfrac{13}{24}$
    (B) 0
    (C) $\dfrac{13\sqrt{3}}{24}$
    (D) $\dfrac{-13}{24}$
    (E) $\dfrac{-13\sqrt{3}}{24}$

3. If $x = \sin t$ and $y = e^{3t}$, what is $\dfrac{dy}{dx}$?

(A) $3e^{3t}$

(B) $\dfrac{3e^{3t}}{\cos t}$

(C) $3e^{3t} \cos t$

(D) $\cos t$

(E) $e^{3t} \cos t$

4. $\lim_{x \to 0} \dfrac{\sin^2(3x)}{x^2}$ is

(A) 0
(B) 1
(C) 3
(D) 9
(E) undefined

5. Where does the function $2x^2 + \sin 2x$ have a point of inflection?

(A) $x = \dfrac{\pi}{4}$

(B) $x = -\dfrac{\pi}{4}$

(C) $x = 0$

(D) $x = \dfrac{\pi}{3}$

(E) $x = -\dfrac{\pi}{3}$

6. A function $f(x)$ is equal to $\dfrac{x^2 - 4}{x - 2}$ for all $x > 0$ except $x = 2$. In order for the function to be continuous at $x = 2$, what must the value of $f(2)$ be?

(A) −4
(B) −2
(C) 0
(D) 2
(E) 4

7. What is $\int_1^4 \sqrt{x}(5x+3)dx$?

   (A) 7.5
   (B) 12
   (C) 64
   (D) 70.4
   (E) 76

8. What is $\lim\limits_{x \to \infty} \dfrac{3\sqrt{x^4 - 3x}}{2x^2 + \cos x}$

   (A) ∞
   (B) undefined
   (C) $\dfrac{3}{2}$
   (D) 0
   (E) 6

9. What is the particular solution to the differential equation $\dfrac{dy}{dx} = \dfrac{x}{\sin y}$ that passes through the point (2, 0)?

   (A) $\cos y = \dfrac{x^2}{2} - 1$
   (B) $\cos y = 3 - \dfrac{x^2}{2}$
   (C) $\cos y = x^2 - 3$
   (D) $\sin y = \dfrac{x^2}{2} - 2$
   (E) $\sin y = 2 - \dfrac{x^2}{2}$

10. Which of the following improper integrals diverge?

(A) $\int_0^{\frac{\pi}{2}} \frac{\sin x}{1-\cos x} dx$

(B) $\int_0^1 \frac{1}{\sqrt{1-x}} dx$

(C) $\int_4^{\infty} xe^{-x^2} dx$

(D) All of the above

(E) None of the above

11. If $f(x) = x\sqrt{2x-1}$, what is $f'(5)$?

(A) 3

(B) $\frac{5}{3}$

(C) $\frac{1}{3}$

(D) $\frac{14}{3}$

(E) $\frac{23}{6}$

12. What is $\lim\limits_{h \to 0} \dfrac{\sin\left(\dfrac{\pi}{2}+h\right)-1}{h}$

(A) $f'\left(\dfrac{\pi}{2}\right)$, where $f(x) = \cos x$

(B) $f'(1)$, where $f(x) = \sin x$

(C) $f'(1)$, where $f(x) = \cos x$

(D) $f'\left(\dfrac{\pi}{2}\right)$, where $f(x) = \sin 2x$

(E) $f'\left(\dfrac{\pi}{2}\right)$, where $f(x) = \sin x$

PRACTICE TEST 1, SECTION I, PART A

13. What is the area under the curve $xe^{(x^2)}$ between $x = 0$ and $x = 2$?

   (A) $\dfrac{e^4}{2} - \dfrac{1}{2}$

   (B) $\dfrac{e^4}{2}$

   (C) $e^4 - 1$

   (D) $4e^4$

   (E) $4e^4 - 4$

14. Which of the following guarantee that $\sum_{n=0}^{\infty} f(n)$ converges?

   (I) $\lim\limits_{x \to \infty} f(x) = 0$

   (II) $f(x) < \dfrac{1}{x^2}$

   (III) $\dfrac{1}{x^2} < f(x)$ when $x \geq 1$.

   (A) (I) only
   (B) (II) only
   (C) (I) and (II) only
   (D) (I) and (III) only
   (E) (I), (II), and (III)

15. What is the absolute maximum value of the function $y = \dfrac{2x}{x^2 + 16}$?

   (A) 4

   (B) $-\dfrac{1}{4}$

   (C) 0

   (D) $\dfrac{1}{4}$

   (E) 4

16. Let $f(x)$ be a function with a continuous derivative on the interval (0, 5) such that
$f'(0) = 3, f'(1) = 2, f'(2) = -3, f'(3) = -4, f'(4) = 1$.
Which of the following must be true about $f(x)$?

   (I) $f(x)$ has a critical point between $x = 1$ and $x = 2$
   (II) $f(x)$ has a critical point between $x = 0$ and $x = 1$
   (III) $f(x)$ has a critical point between $x = 2$ and $x = 3$

   (A) (I) only
   (B) (III) only
   (C) (I) and (II) only
   (D) (I) and (III) only
   (E) (I), (II), and (III)

17. Let f be a continuous function on the interval $[-1, 3]$. If $f(-1) = 9$ and $f(3) = 1$, then the Mean Value Theorem guarantees that

   (A) $f'(0) = 0$
   (B) $f'(c) = -2$ for some c between -1 and 3
   (C) $f'(c) = 2$ for some c between -1 and 3
   (D) $f = 5$ for some c between -1 and 3
   (E) $f < 0$ for all c between -1 and 3

18. A young tree is measured every week. Each week, it is taller than it was the week before, but it has grown a lesser amount. If $h(t)$ is the height of the tree, which of the following is positive?

   (I) $h(t)$
   (II) $h'(t)$
   (III) $h''(t)$

   (A) (II) only
   (B) (I) and (II) only
   (C) (I) and (III) only
   (D) (II) and (III) only
   (E) (I), (II), and (III)

19. On what interval is the function $y = \dfrac{6x}{x^2 + 9}$ increasing?

   (A) $(-\infty, 0)$
   (B) $(-3, 0)$
   (C) $(0, \infty)$
   (D) $(0, 3)$
   (E) $(-3, 3)$

20. Let $f(x) = \dfrac{x}{x+1}$. Which of the following are true?

   (I) $f(x)$ has exactly one local maximum
   (II) $f(x)$ has a point of inflection at $x = 0$
   (III) $f(x)$ has a vertical asymptote at $x = -1$

   (A) (I) only
   (B) (III) only
   (C) (I) and (III) only
   (D) (II) and (III) only
   (E) (I), (II), and (III)

21. What is the radius of convergence of the series $\sum_{n=0}^{\infty} \dfrac{(x-2)^n}{n!}$?

   (A) 0
   (B) 1
   (C) 2
   (D) 3
   (E) ∞

22. What is the volume of the solid obtained by rotating the region between $y = \dfrac{6}{x+1}$ and $y = 4 - x$ around the x-axis?

   (A) $\pi \int_1^3 (4-x)^2 - \left(\dfrac{6}{x+1}\right)^2 dx$

   (B) $\pi \int_1^3 \left(\dfrac{6}{x+1}\right)^2 - (4-x)^2 dx$

   (C) $\pi \int_1^2 (4-x)^2 - \left(\dfrac{6}{x+1}\right)^2 dx$

   (D) $2\pi \int_1^2 (4-x) - \left(\dfrac{6}{x+1}\right) dx$

   (E) $2\pi \int_1^3 (4-x) - \left(\dfrac{6}{x+1}\right) dx$

23. This is a slope field for which of the following differential equations?

(A) $\dfrac{dy}{dx} = xy^2$

(B) $\dfrac{dy}{dx} = xy$

(C) $\dfrac{dy}{dx} = \dfrac{x}{y}$

(D) $\dfrac{dy}{dx} = \dfrac{x^2}{y}$

(E) $\dfrac{dy}{dx} = x^2 y$

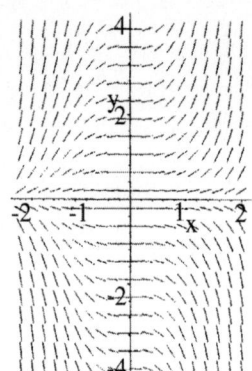

This graph of $f(x)$ is for problems 24 and 25.

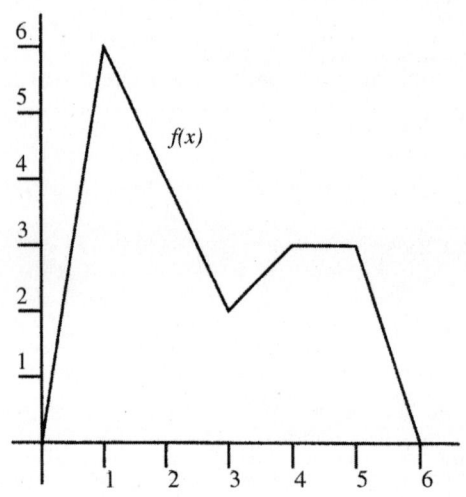

24. Let $g(x) = \int_0^t f(x)\,dx$. What is $g(6)$?

(A) 0
(B) 6
(C) 12
(D) 18
(E) unknown

25. Where is $f'(x) = 0$?

   (I) When $x = 1$
  (II) When x is between 4 and 5
 (III) When $x = 2$

(A) (I) only
(B) (II) only
(C) (I) and (II) only
(D) (II) and (III) only
(E) (I), (II), and (III)

26. What is $\int \frac{9x+10}{2x^2 - x - 6} dx$?

(A) $4\ln|x-2| + \frac{\ln|2x+3|}{2} + C$

(B) $4\ln|9x+10| + \frac{\ln|2x+3|}{2} + C$

(C) $3\ln|x-2| + \ln|2x+3| + C$

(D) $3\ln|9x+10| + 2\ln|2x+3| + C$

(E) $4\ln|x-2| + 2\ln|2x+3| + C$

27. A particle is moving along the x-axis. Its position at time $t > 0$ is e^{2-t}. What is its acceleration when $t = 2$?

(A) e
(B) 1
(C) 0
(D) -1
(E) $-e$

28. A particle, initially at position (1, 2), moves with velocity $(t, \sin(\pi t))$. Where is the object when $t = 2$?

(A) $\left(1, 2 + \dfrac{1}{\pi}\right)$

(B) $\left(3, 2 + \dfrac{1}{\pi}\right)$

(C) $\left(3, 2 - \dfrac{1}{\pi}\right)$

(D) (1, 2)

(E) (3, 2)

PRACTICE TEST 1, SECTION I, PART B

Time	Number of Questions
50 minutes	17

A GRAPHING CALCULATOR IS REQUIRED FOR SOME QUESTIONS
ON THIS PART OF THE EXAMINATION

SECTION I, PART B

In this test: The exact numerical value of the correct answer does not always appear among the choices given. When this happens, select from among the choices the number that best approximates the exact numerical value.

Unless otherwise specified, the domain of a function f is assumed to be the set of all real numbers x for which $f(x)$ is a real number.

29. The position of an object is given by $y = \dfrac{\cos 5t}{6} - \dfrac{\sin 5t}{2}$, where t is the time in seconds. In the first 2 seconds, how many times is the velocity of the object equal to 0?

   (A) 0
   (B) 1
   (C) 3
   (D) 5
   (E) 6

30. Which of the following series converge?

   (A) $\sum\limits_{n=1}^{\infty} \dfrac{(-1)^n}{\ln(n+1)}$

   (B) $\sum\limits_{n=4}^{\infty} \dfrac{\sqrt{n}}{3n^2 + n - 2}$

   (C) $\sum\limits_{n=0}^{\infty} \dfrac{3^n}{(2n)!}$

   (D) All of the above

   (E) None of the above

31. Each edge of a cube is increasing at a rate of 5 cm per minute. Find the rate at which the surface area is increasing when the sides are 10cm long.

   (A) 30cm^2 /minute
   (B) 60cm^2 /minute
   (C) 180cm^2 /minute
   (D) 300cm^2 /minute
   (E) 600cm^2 /minute

32. What is $\int (x+1)\cos x\, dx$?

   (A) $(x+1)\sin x + \cos x + C$
   (B) $\left(\dfrac{x^2}{2} + x\right)\sin x + C$
   (C) $(x+1)\sin x - \cos x + C$
   (D) $(x+1)\cos x + \sin x + C$
   (E) $(x^2 + x)\sin x + \cos x + C$

33. What is the area inside of one loop of the curve $r = 4\sin 3\theta$?

   (A) 1.33
   (B) 2.09
   (C) 4.19
   (D) 8.38
   (E) 12.57

34. Find the maximum area of a right triangle with hypotenuse 7.

   (A) 10.25
   (B) 12.25
   (C) 15.75
   (D) 20.5
   (E) 24.5

35. At what points does the function $x^3 - 3xy + y^2 = 0$ have a horizontal tangent line?

(A) (0, 0) only
(B) (2, 4) only
(C) (2, 2) only
(D) (0, 0) and (2, 4) only
(E) (0, 0) and (2, 2) only

36. For what values of r does $\sum_{n=0}^{\infty} \frac{3r^{2n}}{5^n}$ converge?

(A) $-2.24 < r < 2.24$
(B) $-1.67 < r < 1.67$
(C) $-5 < r < 5$
(D) For all values of r
(E) $r = 0$ only

37. What is the arc length of the curve given by $x = \sin t$, $y = t + \cos t$ as t goes from 0 to $\pi/3$?

(A) 1.04
(B) 1.09
(C) 1.21
(D) 1.41
(E) 1.52

38. A bacteria colony grows at a rate proportional to its size. The population doubles after 3 hours. How long does it take to triple?

(A) 4.5 hours
(B) 4.566 hours
(C) 4.755 hours
(D) 4.897 hours
(E) 5.342 hours

39. If $f(x) = x\ln(\sqrt{x})$, what is $f'(x)$?

(A) $\ln(\sqrt{x}) + \dfrac{1}{2}$

(B) $\ln(\sqrt{x}) + \dfrac{x}{2}$

(C) $\ln(\sqrt{x}) + 1$

(D) $\ln(\sqrt{x}) + \dfrac{\sqrt{x}}{2}$

(E) $\ln\left(\dfrac{1}{2\sqrt{x}}\right)$

40. The motion of a particle in a plane is given by a pair of equations $x(t) = \cos t$, $y(t) = \ln(t+1)$. What is its speed when $t = 3$?

(A) 0.29
(B) 0.39
(C) 0.64
(D) 0.66
(E) 0.82

41. When does the function $f(x) = \sqrt[3]{2x+1}$ have a vertical tangent line?

(A) $x = -1$

(B) $x = -\dfrac{1}{2}$

(C) $x = 0$

(D) $x = \dfrac{1}{2}$

(E) Nowhere

42. If $f(4) = 5$ and $f'(4) = 2$, an estimate of $f(3.9)$ is

(A) 5.2
(B) 5
(C) 4.8
(D) 4
(E) 2

43. The first 4 terms of the Taylor series around 0 of $\dfrac{1}{1+2x}$ are

   (A) $1 - 2x + 4x^2 - 8x^3$
   (B) $1 + 2x + 4x^2 + 8x^3$
   (C) $1 - x + 2x^2 - 4x^3$
   (D) $1 + x + 2x^2 + 4x^3$
   (E) $1 - 2x^2 + 4x^4 - 8x^6$

44. A particle is moving along the x-axis with velocity $e^{3t} - 1$, for $t > 0$. When $t = 0$, it is at the position $x = 1$. What is its position when $t = 2$?

   (A) $e^6 + 4$

   (B) $\dfrac{e^6 - 2}{3}$

   (C) $\dfrac{e^6 - 4}{3}$

   (D) $3e^6 - 1$

   (E) $\dfrac{e^6}{3} - 2$

45. What is the slope of the tangent line to the curve $y = x \tan x$ when $x = \dfrac{\pi}{4}$?

   (A) 1.39
   (B) 1.79
   (C) 2.57
   (D) 2.79
   (E) 3.14

END OF SECTION I

IF YOU FINISH BEFORE TIME IS CALLED, YOU MAY CHECK YOUR WORK ON THIS SECTION.
DO NOT GO TO SECTION II UNTIL YOU ARE TOLD TO DO SO.

ADVANCED PLACEMENT CALCULUS BC

Time	Number of Questions
45 minutes	3

A GRAPHING CALCULATOR IS REQUIRED FOR SOME QUESTIONS
ON THIS PART OF THE EXAMINATION

Section II, Part A

SHOW ALL YOUR WORK. Indicate your methods clearly because you will be graded on the correctness and completeness of your methods as well as the accuracy of your final answers. Correct answers without supporting work may not receive credit.

You are permitted to use your calculator to solve an equation, find the derivative of a function at a point, or calculate the value of a definite integral. However, you must clearly indicate the setup of your program, namely the equation, function, or integral you are using. If you use other built-in features or programs, you must show the mathematical steps necessary to produce your results.

Unless otherwise specified, answers (numeric or algebraic) need not be simplified. If your answer is given as a decimal approximation, it should be correct to three places after the decimal point.

---

1. Let $f(x)$ be a function that has derivatives of all orders for all real numbers. Assume $f(0) = 2$, $f'(0) = 3$, $f''(0) = -1$, and $f'''(0) = 4$.

    (A) Write the third-degree Taylor polynomial for $f(x)$ about $x = 0$ and use it to approximate $f(-0.1)$.
    (B) Write the fourth-degree Taylor polynomial for $g(x)$, where $g(x) = f(-x^2)$, about $x = 0$.
    (C) Write the third-degree Taylor polynomial for $h(x)$, where $h(x) = \int_0^x f(t)\,dt$, about $x = 0$.

---

2. At a certain location, the number of hours of daylight is given by

$$L(t) = 12 + 3\sin\left(\frac{\pi t}{183}\right), \quad 0 \le t < 365$$

    where t is the number of days since the spring equinox.

    (A) Sketch the graph of $L(t)$.
    (B) Find the average number of hours of daylight between $t = 0$ and $t = 91$.
    (C) A building at this location has a solar collector, which turns sunlight into energy. There is excess energy produced whenever there are 13 hours or more of daylight. For what values of t is extra energy collected?
    (D) The extra energy produced from the hours of sunlight after the first 13 can be sold to the local power company at a rate of $0.30 per hour collected. How much money is made by selling the energy in a year?

3. Let R be the region in the plane bounded by the curve $y = (x+1)^{1/2}$, the y-axis, and the line $y = 2$.

   (A) Find the area of the region R.
   (B) Find the volume of the solid generated when R is revolved around the x-axis.
   (C) The vertical line $x = k$ divides the region R into two solids such that when these two solids are revolved around the x-axis, they generate solids with equal volume. Find the value of k.

ADVANCED PLACEMENT CALCULUS BC

Time	Number of Questions
45 minutes	3

SECTION II, PART B

You may want to look over the problems before starting to work on them, since it is not expected that everyone will be able to complete all parts of all problems. All problems are given equal weight, but the parts of a particular problem are not necessarily given equal weight.

A GRAPHING CALCULATOR MAY NOT BE USED ON THIS PART OF THE EXAMINATION

You may continue to work on Part A, but you may not use your calculator.

SHOW ALL YOUR WORK. Indicate your methods clearly because you will be graded on the correctness and completeness of your methods as well as the accuracy of your final answers. Correct answers without supporting work may not receive credit.

Unless otherwise specified, answers (numeric or algebraic) need not be simplified.

---

4. The velocity of a particle moving along the x-axis is given in the following table and graph.

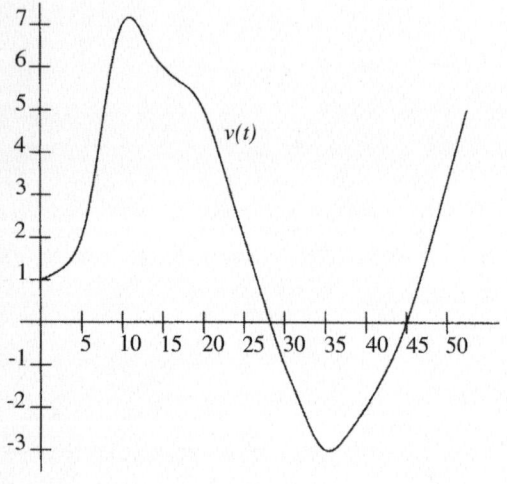

t	$v(t)$
0	1
5	2
10	7
15	6
20	5
25	2
30	−1
35	−3
40	−2
45	0
50	5

The position of the particle at time t is given by $x(t)$, $0 \le t \le 50$.

(A) The particle is at position $x = 10$ when $t = 0$. Estimate the position of the particle when $t = 20$.
(B) When is $x(t)$ smallest? Justify your answer.
(C) Approximate the acceleration of the particle when $t = 30$.
(D) When is the acceleration of the particle positive?

5. Consider a particle moving in the plane with $\dfrac{dx}{dt} = e^{-3t}$ and $\dfrac{dy}{dt} = \dfrac{1}{(t+1)^2}$, for $t > 0$, which is at the point (5/3, 2) when $t = 0$.

(A) Where is the particle when $t = 3$?
(B) What is the speed of the particle when $t = 1$?
(C) What point does the particle approach as t approaches infinity?

6. Consider the differential equation $\dfrac{dy}{dx} = \dfrac{(x+y)}{4}$

(A) On the axes provided, sketch the slope field of the given differential equation.

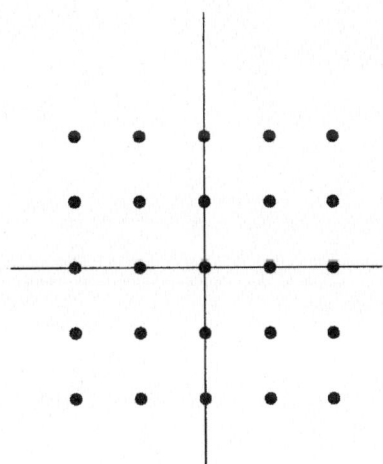

(B) Let $y = f(x)$ be the particular solution to the differential equation subject to the initial condition $f(0) = 4$. Use Euler's method, starting at $x = 0$, with a step size of 0.1, to approximate $f(0.2)$. Show the work that leads to the answer.
(C) At what points are solutions to this differential equation concave up?

ADVANCED PLACEMENT CALCULUS BC

PRACTICE TEST 2

Time	Number of Questions
55 Minutes	28

A CALCULATOR MAY NOT BE USED ON THIS PART OF THE EXAMINATION

Section I, Part A

Directions: Solve each of the following problems using the available space for scratchwork. After examining the form of the choices, decide which is the best of the choices given and fill in the corresponding oval on the answer sheet. No credit will be given for anything written in the test book. Do not spend too much time on any one problem.

1. At which of the following points is the function e^{1-x^2} concave down?

    (A) $\dfrac{1}{2}$

    (B) $\dfrac{3}{2}$

    (C) 1

    (D) $-\dfrac{3}{2}$

    (E) $-\dfrac{3}{4}$

2. A spherical balloon is being blown up, with air entering it at a constant rate. Let $r(t)$ be the radius of the balloon. Which of the following is positive?

   (I) $r(t)$
   (II) $r'(t)$
   (III) $r''(t)$

   (A) (I) only
   (B) (II) only
   (C) (III) only
   (D) (I) and (II) only
   (E) (I),(II), and (III)

3. This is a slope field for which of the following differential equations?

   (A) $\dfrac{dy}{dx} = xy$

   (B) $\dfrac{dy}{dx} = \dfrac{x^2}{y}$

   (C) $\dfrac{dy}{dx} = x^2 y$

   (D) $\dfrac{dy}{dx} = \dfrac{y}{x}$

   (E) $\dfrac{dy}{dx} = \dfrac{x}{y}$

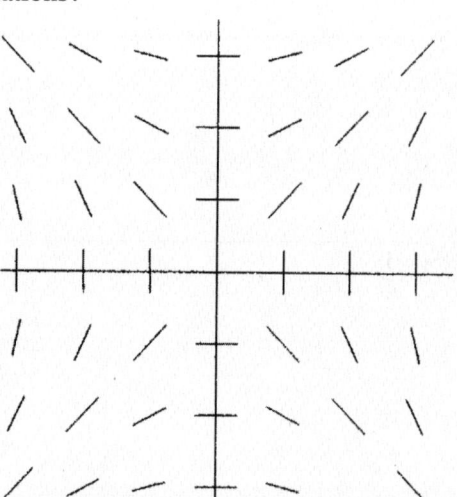

4. What is the radius of convergence of the series $\displaystyle\sum_{n=0}^{\infty} \dfrac{x^n}{2\ln n}$

   (A) 0
   (B) 1
   (C) 2
   (D) e
   (E) ∞

PRACTICE TEST 2, SECTION I, PART A

5. What is the particular solution to the differential equation $\frac{dy}{dx} = (3x^2 - 1)y$ that passes through the point (2, 1)?

   (A) $y = e^{x^3 - x - 6}$

   (B) $y^2 = 2x^3 - 2x - 11$

   (C) $y = e^{3x^2 - 1}$

   (D) $y = 3x^2 - 7$

   (E) $y = e^{x^3 - x}$

6. What is $\sum_{n=0}^{\infty} \frac{3^n}{2^{2n}}$?

   (A) 0

   (B) 1

   (C) $\frac{3}{2}$

   (D) 4

   (E) Does not converge

7. A particle which is at position (1, 3) with velocity (0, −1) when $t = 0$, moves with acceleration $\left(1, \frac{1}{(t+1)^2}\right)$. Where is the object when $t = 2$?

   (A) (3, 3 − ln 3)
   (B) (2, ln 3)
   (C) (2, 3 − ln 3)
   (D) (3, − ln 3)
   (E) (3, 3 + ln 3)

This graph of $f'(x)$ is for problems 8 and 9.

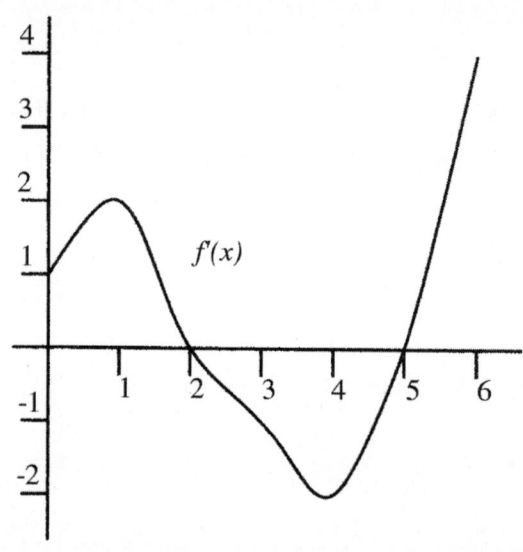

8. Where does $f(x)$ have a local maximum on the interval $[0, 6]$?

   (A) $x = 0$
   (B) $x = 1$
   (C) $x = 2$
   (D) $x = 4$
   (E) $x = 5$
   (F) nowhere

9. Which of the following are true about the graph of $f''(x)$?

   (I) It is decreasing on $(0, 1)$
   (II) It is positive on $(4, 5)$
   (III) It has a local minimum when $x = 5$

   (A) (I) only
   (B) (II) only
   (C) (I) and (II) only
   (D) (II) and (III) only
   (E) (I), (II), and (III)

10. Let R be the region between the curves $y = \sqrt{x^3 + 1}$ and $y = x + 1$, for which x is positive. What is the volume of the solid obtained by rotating R around the x-axis?

(A) $\pi \int_0^3 (x^3 - x^2 - 2x - 2)dx$

(B) $\pi \int_0^2 (-x^3 + x^2 + 2x)dx$

(C) $\pi \int_0^2 (x^3 - x^2 - 2x - 2)dx$

(D) $\pi \int_1^3 (x^3 - x^2 - 2x + 2)dx$

(E) $\pi \int_1^2 (-x^3 + x^2 + 2x + 2)dx$

11. This is the graph of which function?

(A) $\dfrac{x}{e^x}$

(B) $\dfrac{x^2}{e^x}$

(C) $\dfrac{e^x}{x}$

(D) $x^2 e^x$

(E) xe^x

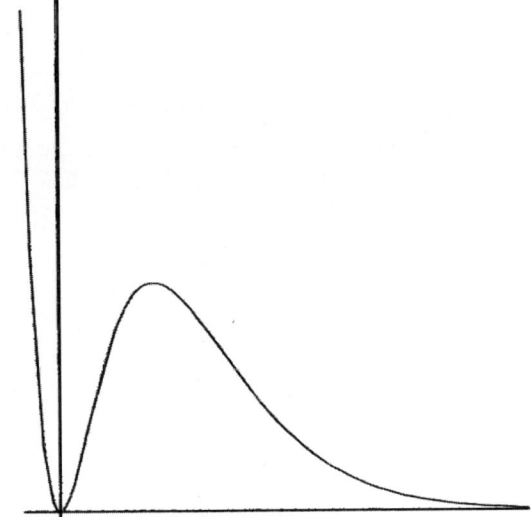

12. Let $f(x) = \dfrac{2}{x^2+4}$ Which of the following are true?

   (I) $f(x)$ has an absolute maximum at $x = 0$

   (II) $f(x)$ has a vertical asyptote at $x = -2$

   (III) $f(x)$ has a horizontal asymptote at $y = \dfrac{1}{2}$

   (A) (I) only
   (B) (II) only
   (C) (I) and (II) only
   (D) (II) and (III) only
   (E) (I), (II), and (III)

13. If $x = \sqrt{t}$ and $y = \sin 4t$, what is $\dfrac{dy}{dx}$?

   (A) $\dfrac{2\sqrt{t}}{\cos 4t}$

   (B) $2\sqrt{t}\cos 4t$

   (C) $\dfrac{2\cos 4t}{\sqrt{t}}$

   (D) $8\sqrt{t}\cos 4t$

   (E) $\dfrac{8\cos 4t}{\sqrt{t}}$

14. A particle is moving along the x-axis with velocity $3t^2 - t$, for $t > 0$. When $t = 0$, it is at the position $x = 3$. What is its position when $t = 2$?

   (A) 6
   (B) 9
   (C) 10
   (D) 13
   (E) 24

15. A function $f(x)$ is equal to $\dfrac{\sin 2x}{x}$ for all x except $x = 0$. In order for the function to be continuous at $x = 0$, what must the value of $f(0)$ be?

   (A) 2
   (B) 1
   (C) 0
   (D) −1
   (E) −2

16. What is $\lim\limits_{x \to \infty} \dfrac{\sqrt{3x^2 + 2x + 1}}{x+1}$

   (A) 3
   (B) ∞
   (C) $\sqrt{3}$
   (D) 0
   (E) $\dfrac{\sqrt{3}}{2}$

17. If $f(x) = \cos(x-1)e^{x^2-1}$, what is $f'(1)$?

   (A) 2
   (B) 1
   (C) −1
   (D) −2
   (E) 0

18. What is the slope of the tangent line to the curve $y = \dfrac{1}{\sqrt{x^2-5}}$ when $x = 3$?

   (A) $-\dfrac{3}{8}$

   (B) $\dfrac{3}{2}$

   (C) $\dfrac{1}{2}$

   (D) $-\dfrac{1}{4}$

   (E) $-\dfrac{3}{16}$

19. What is $\int_1^2 \dfrac{1}{3x+1}\,dx$?

   (A) $-\dfrac{3}{28}$

   (B) $3\ln\left(\dfrac{7}{4}\right)$

   (C) $\dfrac{(\ln 7)}{3}$

   (D) $\dfrac{(\ln 7)}{(3\ln 4)}$

   (E) $\dfrac{\ln\left(\dfrac{7}{4}\right)}{3}$

20. What is $\int \dfrac{5x+3}{3x^2-2x-1}dx$?

(A) $2\ln|x-1| - 3\ln|3x+1| + C$

(B) $3\ln|3x-1| + C$

(C) $2\ln|x-1| - \dfrac{\ln|3x+1|}{3} + C$

(D) $\ln|5x+1| + 2\ln|x-1| - \dfrac{\ln|3x+1|}{3} + C$

(E) $3\ln|x-1| + C$

21. On what interval is the function $\dfrac{e^x}{x+1}$ decreasing?

(A) $(0, \infty)$
(B) $(-1, 0)$
(C) $(0, 1)$
(D) $(-\infty, -1)$
(E) $(-\infty, 0)$

22. What is the minimum value of the function $\dfrac{x^2+4}{x}$ on the interval $(1, 3)$?

(A) -4
(B) 4
(C) $\dfrac{13}{3}$
(D) -3
(E) $-\dfrac{13}{3}$

23. What is $\int xe^{2x} dx$?

(A) $e^{2x}\left(\dfrac{1}{4} - \dfrac{x}{2}\right) + C$

(B) $e^{2x}\left(\dfrac{x}{4} - \dfrac{1}{2}\right) + C$

(C) $e^{2x}\left(\dfrac{x}{2} - \dfrac{1}{4}\right) + C$

(D) $e^{2x}\left(\dfrac{1}{2} - \dfrac{x^2}{4}\right) + C$

(E) $\dfrac{e^{2x} x^2}{2} - \dfrac{1}{4} + C$

24. Let f be a continuous function on the interval $[-1, 9]$. If $f(-1) = 2$ and $f(9) = 7$, then which of the following are necessarily true?

  (I) $f'(c) = \dfrac{1}{2}$ for some c between -1 and 9
  (II) $f(c) > 0$ for all c between -1 and 9
  (III) $f(c) = 5$ for some c between -1 and 9

(A) (I) only
(B) (II) only
(C) (I) and (II) only
(D) (I) and (III) only
(E) (I), (II), and (III)

25. When does the function $f(x) = e^{\sqrt[3]{x-1}}$ have a vertical tangent line?

(A) $x = 1$
(B) $x = 0$
(C) $x = -1$
(D) $x = \dfrac{1}{3}$
(E) Nowhere

26. What is $\lim_{a \to 4} \dfrac{2-\sqrt{a}}{4-a}$?

(A) $f'(2)$, where $f(x) = \sqrt{x}$

(B) $f'(2)$, where $f(x) = 1 - \sqrt{x}$

(C) $f'(4)$, where $f(x) = \sqrt{x}$

(D) $f'(2)$, where $f(x) = \sqrt{4-x}$

(E) $f'(4)$, where $f(x) = \dfrac{1}{\sqrt{4}}$

27. Which of the following guarantee that $\sum_{n=0}^{\infty} f(n)$ converges?

(I) $\lim_{n \to \infty} \dfrac{f(n+1)}{f(n)} = 1$

(II) $f(n) = (-1)^n g(n)$, where $g(x)$ is a positive function that tends to 0 as x tends to infinity.

(III) $0 < f(x) < h(x)$ for $0 < x$ and $\sum_{n=0}^{\infty} h(n)$ converges.

(A) (I) only
(B) (II) only
(C) (II) and (III) only
(D) (I) and (III) only
(E) (I), (II), and (III)

28. Let $f(x)$ be a function such that $\int_1^3 f(x)dx = 1$ and $\int_3^4 f(x)dx = 2$. Let $g(x)$ be a function such that $\int_1^4 g(x)dx = 3$. What is $\int_1^4 \dfrac{2f(x)}{g(x)} dx$?

(A) 0
(B) 1
(C) 2
(D) 3
(E) insufficient information to determine the answer

ADVANCED PLACEMENT CALCULUS BC

Time	Number of Questions
50 Minutes	17

A GRAPHING CALCULATOR IS REQUIRED FOR SOME QUESTIONS ON THIS PART OF THE EXAMINATION.

SECTION I, PART B

Directions: Solve each of the following problems using the available space for scratchwork. After examining the form of the choices, decide which is the best of the choices given and fill in the corresponding oval on the answer sheet. No credit will be given for anything written in the test book. Do not spend too much time on any one problem.

In this test: The exact numerical value of the correct answer does not always appear among the choices given. When this happens, select from among the choices the number that best approximates the exact numerical value.

Note: Unless otherwise specified, the domain of a function f is assumed to be the set of all real numbers x for which $f(x)$ is a real number.

29. Consider the series $\sum_{n=0}^{\infty} \dfrac{(-1)^n}{2\ln(n+1)}$. To estimate the sum of this series within 0.1 of the correct answer, how many terms must be added?

   (A) 11
   (B) 96
   (C) 101
   (D) 135
   (E) 148

30. If $f(x) = \dfrac{e^{\ln(\sin x)}}{\cos x}$, what is $f'(x)$?

   (A) $e^{\ln(\sin x)} \sin x \sec^2 x$
   (B) $\sec^2 x$
   (C) $\cos^2 x$
   (D) $\sin x e^{\ln(\cos x)}$
   (E) $e^{\ln(\sin x)} \sec^2 x$

31. A particle is moving along the x-axis. Its position at time $t > 0$ is $\ln(2t^{\frac{3}{2}} + 1)$. What is its speed when $t = 4$?

   (A) 2.01
   (B) 3.06
   (C) 0.353
   (D) 4.63
   (E) 7.81

32. If $f(2) = 3.15$ and $f'(2) = -0.5$, an estimate of $f(2.1)$ is

   (A) 3
   (B) 3.05
   (C) 3.1
   (D) 3.2
   (E) 3.15

33. The first 4 terms of the Taylor series around 0 of $\cos(\sqrt{x})$ are

   (A) $1 - \dfrac{x}{2} + \dfrac{x^2}{4!} - \dfrac{x^3}{6!}$

   (B) $1 + \dfrac{x}{2} + \dfrac{x^2}{4!} + \dfrac{x^3}{6!}$

   (C) $1 - \dfrac{x}{2} + \dfrac{x^2}{4} + \dfrac{x^3}{6}$

   (D) $1 + \dfrac{x}{2} + \dfrac{x^2}{4} + \dfrac{x^3}{6}$

   (E) $1 - \dfrac{x}{2} - \dfrac{x^2}{4!} - \dfrac{x^3}{6!}$

34. What is the area under the curve $\sec^2(4x)$, as x goes from 0 to t?

   (A) $4 \tan 4t$

   (B) $\dfrac{\tan 4t}{4}$

   (C) $\tan 4t - 1$

   (D) $\dfrac{\tan 4t}{4} - 1$

   (E) $4 \tan 4t - 1$

35. What is $\lim_{x \to \infty} \dfrac{2x^{\frac{2}{3}} + \sin x}{x - 3}$?

   (A) ∞
   (B) undefined
   (C) 0
   (D) 1
   (E) -3

36. Where does the function $\ln(x^2 + 2x + 2)$ have a point of inflection?

   (A) At $x = -2$ and $x = 0$ only
   (B) At $x = -1$ and $x = 1$ only
   (C) At $x = 0$ only
   (D) At $x = -2$ only
   (E) At $x = -2$, $x = 0$, and $x = 1$

37. A conical funnel of height 4cm and radius 2cm is full of water. The water is pouring out at a rate of 1cm³ per second. If $h(t)$ is the height of the water level in the cone, what is $h'(t)$ when $h(t) = 2$?

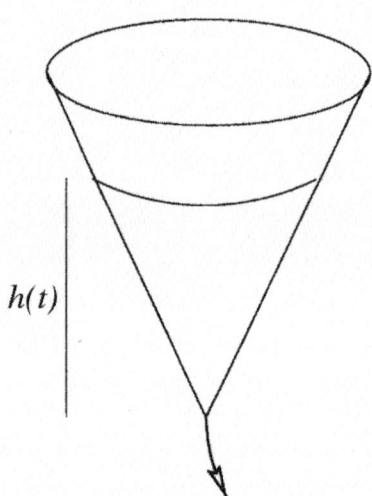

   (A) -0.51 cm/sec
   (B) -0.44 cm/sec
   (C) -0.37 cm/sec
   (D) -0.32 cm/sec
   (E) -0.28 cm/sec

38. For what values of r does $\sum_{n=0}^{\infty} \dfrac{n^2+1}{(n+1)^r}$ converge?

   (A) $r > 2$
   (B) $r < 3$
   (C) $r > 3$
   (D) $r = 0$
   (E) Never

39. Find the maximum area of a rectangle with perimeter 12.

   (A) 9
   (B) 12
   (C) 15
   (D) 16
   (E) 36

40. Which of the following improper integrals converge?

   (A) $\int_1^5 \dfrac{1}{1-x} dx$

   (B) $\int_1^{\infty} \sin x \, dx$

   (C) $\int_1^{\infty} \dfrac{e^{-\sqrt{x}}}{\sqrt{x}} dx$

   (D) None of the above

41. What is the area inside the curve $r = 2\cos\theta$ and outside the curve $r = 1$?

   (A) 1.37
   (B) 1.52
   (C) 1.76
   (D) 1.84
   (E) 1.92

42. Which of the following series converge?

(A) $\sum_{n=1}^{\infty} \frac{(-1)^n}{2n+1}$

(B) $\sum_{n=0}^{\infty} \frac{1}{\sqrt{3n+4}}$

(C) $\sum_{n=0}^{\infty} \frac{2n}{\sqrt{n+1}}$

(D) All of the above

(E) None of the above

43. What is the arc length of the curve given by $x = t^2 + 1$, $y = \frac{1}{t}$ as t goes from 1 to 2?

(A) 1.50
(B) 3.09
(C) 3.69
(D) 5.17
(E) 9.83

44. A radioactive substance has a half-life of 20 minutes. What percentage is left after 11 minutes?

(A) 53%
(B) 59%
(C) 61%
(D) 68%
(E) 72%

45. The motion of a particle in a plane is given by a pair of equations $x(t) = \sqrt{t+1}$, $y(t) = e^{1-t}$. What is its speed when $t = 2$?

(A) 0.22
(B) 0.47
(C) 1.05
(D) 1.54
(E) 2.73

PRACTICE TEST 2, SECTION II, PART A

Time	Number of Questions
45 Minutes	3

A GRAPHING CALCULATOR IS REQUIRED FOR SOME QUESTIONS ON THIS PART OF THE EXAMINATION.

Section II, Part A

Directions: Solve each of the following problems, using the available space for scratchwork. After examining the form of the choices, decide which is the best of the choices given and fill in the corresponding oval on the answer sheet. No credit will be given for anything written in the test book. Do not spend too much time on any one problem.

In this test: SHOW ALL YOUR WORK. Indicate your methods clearly because you will be graded on the correctness and completeness of your methods as well as the accuracy of your final answers. Correct answers without supporting work may not receive credit.

You are permitted to use your calculator to solve an equation, find the derivative of a function at a point, or calculate the value of a definite integral. However, you must clearly indicate the setup of your program, namely the equation, function, or integral you are using. If you use other built-in features or programs, you must show the mathematical steps necessary to produce your results.

Note: Unless otherwise specified, answers (numeric or algebraic) need not be simplified. If your answer is given as a decimal approximation, it should be correct to three places after the decimal point.

---

1. A rumor is spreading through a population at a rate of $r(t) = \dfrac{1000}{1+(t-10)^2}$ people per day.

   (A) When is the rumor spreading the most quickly? Justify your answer.
   (B) If 5 people know the rumor when $t = 0$, how many people have heard it when $t = 40$?
   (C) As t approaches infinity, what does the number of people who have heard the rumor approach?

---

2. Let R be the region in the plane bounded by the curve $y = \dfrac{1}{x+3}$ and the x-axis, between $x = 0$ and $x = 1$.

   (A) Find the area of the region R.
   (B) Find the volume of the solid generated when R is revolved around the x-axis.
   (C) Consider the region T bounded by the curve $y = \dfrac{1}{kx+3}$ and the x-axis, between $x = 0$ and $x = 1$. What is the value of k such that the volume of the solid generated by revolving T around the x-axis is twice that obtained by revolving R around the x-axis?

3. A particle is moving along the curve $y = x^2 - x + 1$ with a speed of 3 units per second, its x-coordinate is increasing. The particle starts at the point $(-2, 7)$. Let t be the number of seconds since it left that point.

   (A) How long does it take the particle to move to the point $(0, 1)$?
   (B) What is the velocity of the particle when $x = 1$?
   (C) When $x = 1$, is the x-coordinate of the velocity of the particle increasing or decreasing? Justify your answer by appealing to the graph of the function

PRACTICE TEST 2, SECTION II, PART B

Time	Number of Questions
45 Minutes	3

A GRAPHING CALCULATOR MAY NOT BE USED ON THIS PART OF THE EXAMINATION

Section II, Part B

Directions: Solve each of the following problems using the available space for scratchwork. After examining the form of the choices, decide which is the best of the choices given and fill in the corresponding oval on the answer sheet. No credit will be given for anything written in the test book. Do not spend too much time on any one problem.

<u>In this test:</u> You may want to look over the problems before starting to work on them, since it is not expected that everyone will be able to complete all parts of all problems. All problems are given equal weight, but the parts of a particular problem are not necessarily given equal weight.

You may continue to work on Part A, but you may not use your calculator.

Show all your work. Indicate your methods clearly because you will be graded on the correctness and completeness of your methods as well as the accuracy of your final answers. Correct answers without supporting work may not receive credit.

<u>Note:</u> Unless otherwise specified, answers (numeric or algebraic) need not be simplified.

4. The velocity of a particle moving along the x-axis at t goes from 0 to 18 is given in the following table and graph:

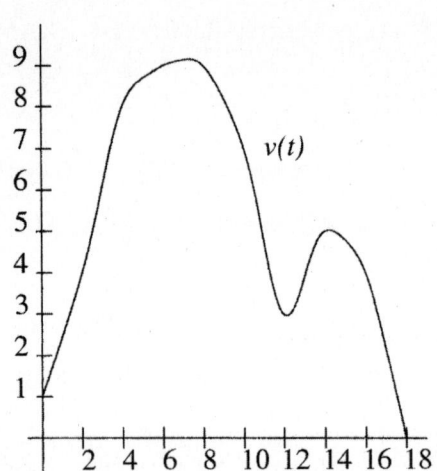

t	$v(t)$
0	1
2	4
4	8
6	9
8	9
10	7
12	3
14	5
16	4
18	0

(A) Estimate the acceleration of the particle when $t = 10$.
(B) What is the average acceleration of the particle, from $t = 0$ to $t = 18$?
(C) When is the particle farthest from its starting position? Justify your answer.
(D) Estimate the total distance traveled by the particle.

5. Consider the differential equation $\dfrac{dy}{dx} = \dfrac{x^2}{y}$.

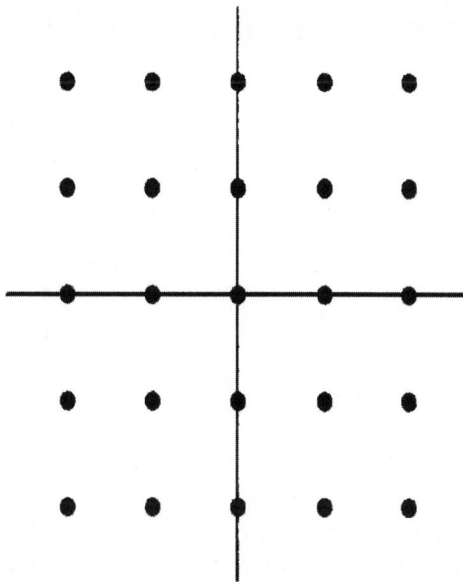

(A) On the axes provides, sketch the slope field of the given differential equation.
(B) Let $y = f(x)$ be the particular solution to the differential equation subject to the initial condition $f(0) = -1$. Use Euler's method, starting at $x = 0$, with a step size of 0.1, to approximate $f(0.2)$. Show the work that leads to the answer.
(C) Find the particular solution to the differential equation with the initial condition $f(0) = -1$.

6. Let $f(x) = e^{-x^2}$.

(A) Write the fourth-degree Taylor polynomial for $f(x)$ about $x = 0$.
(B) Write the fourth-degree Taylor polynomial for $g(x)$, where $g(x) = f(2x)$, about $x = 0$.

(C) Write the fifth-degree Taylor polynomial for $h(x)$, where $h(x) = \int_0^x f(t)\,dt$, about $x = 0$.

ADVANCED PLACEMENT CALCULUS AB

PRACTICE TEST 1: ANSWERS AND EXPLANATIONS

Section I, Part A

1. **The correct answer is (B).**

$$f'(x) = 6 \times 3/2 \times x^{1/2} = 9x^{1/2}$$
$$f''(x) = 9 \times 1/2 \times x^{-1/2} = 9/2 x^{-1/2} = 9/(2x^{1/2}) = 9/(2\sqrt{x})$$
$$f''(16) = 9/(2 \times \sqrt{16}) = 9/(2 \times 4) = 9/8$$

2. **The correct answer is (D).** The slope of a curve is found by finding the first derivative.

$$9x - 4x \ln y = 3$$
$$9 - (4x \times y'/y + 4 \ln y) = 0$$
$$9 - 4xy'/y - 4 \ln y = 0$$
$$-4xy'/y = 4 \ln y - 9$$
$$-4xy' = 4y \ln y - 9y$$
$$y' = (4y \ln y - 9y)/(-4x)$$
$$y'(1/3,1) = (4 \times \ln 1 - 9 \times 1)/(-4 \times 1/3)$$
$$= (4 \times 0 - 9)/(-4/3) = (0 - 9)/(-4/3)$$
$$= -9/(-4/3) = -9 \times -3/4 = 27/4$$

3. **The correct answer is (D).** A relative minimum occurs where the graphs changes from a decreasing function to an increasing function.

$$g(x) = 8x^3 - 3x^2 + 4$$
$$g'(x) = 24x^2 - 6x = 0$$
$$6x(4x - 1) = 0$$
$$x = 0, x = 1/4$$

```
      -1        1/8         1
<-----|---------|-----------|----->
 g'(-1) = 30  g'(1/8) =-(3/8)  g'(1) = 18
 Increasing   Decreasing    Increasing
```

So, the graph changes from decreasing to increasing at $x = 1/4$. The function value at $x = 1/4$ is 63/16. The point, then, is (1/4, 63/16).

4. **The correct answer is (A).**

$$\int x\sqrt{(3x^2+4)}\,dx$$
$$u = 3x^2 + 4$$
$$du = 6x\,dx$$
$$du/6 = x\,dx$$

$$(1/6)\int u^{1/2}\,du = (1/6)(2/3)u^{3/2} + C = (1/9)u^{3/2} + C = (1/9)(3x^2+4)^{3/2} + C$$

5. **The correct answer is (B).** The fundamental period of a sin curve is found by dividing 2π by B where the standard sin curve is expressed by $y = A\sin(Bx + C) + D$. Therefore, since this equation can be written as $y = 6\sin(3\pi x - \pi) + 4$, B is 3π. The fundamental period, then, is $2\pi/3\pi$. The fundamental period is 2/3.

6. **The correct answer is (E).** The easiest way to take the limit as x approaches infinity is to simply take the highest powered term on the top and divide it by the highest powered term on the bottom. This gives us $3x^2/6x^2$, which reduces to 1/2.

7. **The correct answer is (B).** This function can be written $f(x) = [(2x - 5)(x + 3)]/(x + 3)]$. Canceling the $(x + 3)$ factors on the top and the bottom, the simplified function is $f(x) = 2x - 5$. At this point, substitute in -3 for x to get $f(-3) = -11$.

8. **The correct answer is (E).** This function requires the quotient rule.

$$h(x) = \frac{(6x - 3x^2)}{(4 + x)}$$

$$h'(x) = \frac{(4+x)(6-6x) - (6x - 3x^2)(1)}{(4+x)^2}$$

$$= \frac{24 - 18x - 6x^2 - 6x + 3x^2}{(4+x)^2}$$

$$= \frac{-3x^2 - 24x + 24}{(4+x)^2} = \frac{-3(x^2 + 8x - 8)}{(4+x)^2}$$

Although this factorization can be done, it does not allow reducing. Nonetheless, it must be done in order to find a correct answer choice.

PRACTICE TEST 1: ANSWERS AND EXPLANATIONS—SECTION I, PART A

9. **The correct answer choice is (E).** This requires the use of the trigonometric integration of $dx/(1 + x^2)$, which is $\tan^{-1}(x)$. The 6 is a constant. We have $6[\tan^{-1}(1) - \tan^{-1}(-1)] = 6[\pi/4 - (-\pi/4)] = 6[\pi/4 + \pi/4] = 6[\pi/2] = 3\pi$.

10. **The correct answer is (A).**

$$g(x) = \csc x - \cot x$$
$$g'(x) = -\csc x \cot x - (-\csc^2 x) = -\csc x \cot x + \csc^2 x$$
$$g'(\pi/6) = -\csc(\pi/6)\cot(\pi/6) + \csc^2(\pi/6) = -(2)(\sqrt{3}) + (2)^2$$
$$= -2\sqrt{3} + 4 = 4 - 2\sqrt{3}$$

11. **The correct answer is (D).**

$$s(x) = \sin^2 x$$
$$s'(x) = 2\sin x \cos x \text{ (Remember to use the chain rule!)}$$
$$s''(x) = 2(\sin x \times -\sin x + \cos x \times \cos x) = 2(-\sin^2 x + \cos^2 x)$$
$$= -2\sin^2 x + 2\cos^2 x = 2\cos^2 x - 2\sin^2 x$$

12. **The correct answer is (C).**

$$g(-1) = -2$$
$$f(-2) = -5/9$$

13. **The correct answer is (D).**

$$[1/(1 + 1)]\int(3x + 1)^2 dx$$

$$(1/2)\int_{-1}^{1} (3x+1)^2 dx$$

$$u = 3x + 1$$
$$du = 3dx$$
$$(1/3)du = dx$$

$$(1/2)(1/3)\int_{-2}^{4} u^2 du = (1/6)(1/3)u^3 \Big|_{-2}^{4}$$

(Remember to change the limits of integration!)

$$(1/18)[4^3 - (-2)^3] = (1/18)[64 + 8] = (1/18)[72] = 4$$

14. **The correct answer is (D).**

$$g(x) = 2^{4x}$$
$$g'(x) = 2^{4x} \times \ln 2 \times 4$$
$$g''(x) = 4\ln 2 \times 2^{4x} \times \ln 2 \times 4 = 16 \times (\ln 2)^2 \times 2^{4x} = 16 \times (\ln 2)^2 \times (2^4)^x$$
$$= 16 \times (\ln 2)^2 \times 16^x = 16^{1+x} \times (\ln 2)^2 = 16^{x+1}(\ln 2)^2$$

15. **The correct answer is (A).**

$$\lim_{x \to 0} \frac{-\cos x + 1 - \sin x}{x} = \lim_{x \to 0} \frac{(-\cos x + 1) - \sin x}{x}$$
$$= \lim_{x \to 0} \frac{(1 - \cos x)}{x} - \lim_{x \to 0} \frac{\sin x}{x}$$
$$= 0 - 1 = -1$$

16. **The correct answer is (B).**

$$y' = \frac{4x}{y}$$
$$\frac{dy}{dx} = \frac{4x}{y}$$

Separate the variables to get $y\, dy = 4x\, dx$. Integrate both sides of the equation to get $\frac{y^2}{2} = \frac{4x^2}{2} + C_1$, $y^2 = 4x^2 + C_2$. Now, substitute -1 for x and 4 for y. We have $C_2 = 12$. Hence, $y^2 = 4x^2 + 12$. Then, when $y = 6$, x will equal $\pm\sqrt{6}$. Only $-\sqrt{6}$ is given as a choice.

17. **The correct answer is (E).**

$$f(x) = \sin x + \cos x$$
$$f'(x) = \cos x - \sin x = 0$$
$$\cos x = \sin x$$
$$x = \pi/4 \text{ and } x = 5\pi/4$$

However, $5\pi/4$ is not in the interval.

PRACTICE TEST 1: ANSWERS AND EXPLANATIONS—SECTION I, PART A

$$f(\pi/4) = \sqrt{2}$$
$$f(0) = 1$$
$$f(\pi) = -1$$

The absolute maximum function value is $\sqrt{2}$.

18. **The correct answer is (E).** Using integration by parts:

$$\int x^2 e^x dx \to x^2 e^x - \int 2xe^x dx$$
$$u = x^2 \quad dv = e^x dx$$
$$du = 2x\,dx \quad v = e^x$$
$$-2\int xe^x dx \to -2\left[xe^x - \int e^x dx\right]$$
$$u = x \quad dv = e^x dx = xe^x - e^x$$
$$du = dx \quad v = e^x$$

so
$$\int x^2 e^x dx = x^2 e^x - 2\left[xe^x - e^x\right]$$
$$= e^x\left[x^2 - 2x + 2\right] + C$$

19. **The correct answer is (B).**

$$R(x) = 5{,}250x - 10x^2$$
$$MR(x) = R'(x) = 5{,}250 - 20x$$
$$MR(21) = R'(21) = 5{,}250 - 20(21) = 4{,}830$$

20. **The correct answer is (E).**

$$y = \cos^2(3x)$$
$$y' = 2\cos(3x) \times [-\sin(3x)] \times 3 = -6\cos(3x)\sin(3x)$$
$$y'(\pi/4) = -6\cos\left(3\pi/4\right)\cdot\sin\left(3\pi/4\right) = -6\left(-\sqrt{2}/2\right)\left(\sqrt{2}/2\right) = 3$$

21. **The correct answer is (A).**

$$y = \ln(x^2 - 5)$$
$$y' = 2x/(x^2 - 5)$$
$$y'' = \left[(x^2 - 5)(2) - (2x)(2x)\right]/(x^2 - 5)^2 = (2x^2 - 10 - 4x^2)/(x^2 - 5)^2$$
$$= (-2x^2 - 10)/(x^2 - 5)^2$$
$$y''(4) = (-2 \times 4^2 - 10)/(4^2 - 5)^2 = (-2 \times 16 - 10)/(16 - 5)^2$$
$$= (-32 - 10)/11^2 = -42/121$$

22. **The correct answer is (C).**

$$f(x) = \frac{5}{x^2 - 2x - 15} = \frac{5}{(x-5)(x+3)}$$

This is discontinuous when the denominator is 0. This occurs when $x = 5$ and when $x = -3$.

23. **The correct answer is (A).**

$$\ln 2 + \ln 5 - \ln 8 - \ln 15 = (\ln 2 + \ln 5) - (\ln 8 + \ln 15) = (\ln 2 \times 5) - (\ln 8 \times 15)$$
$$= \ln 10 - \ln 120 = \ln(10/120) = \ln(1/12)$$
$$= \ln 1 - \ln 12 = 0 - \ln 12 = -\ln 12$$

24. **The correct answer is (C).**

$\sin^2 x + 4 - \cos^2 x = 4 - \sin^2 x - \cos^2 x = 4 - (\sin^2 x + \cos^2 x) = 4 - 1 = 3$

25. **The correct answer is (B).**

$$y = \sqrt{x^3} = x^{3/2}$$
$$y' = (3/2)x^{1/2}$$
$$y'' = (3/2)(1/2)x^{-1/2} = (3/4)x^{-1/2}$$
$$y''' = (3/4)(-1/2)x^{-3/2} = (-3/8)x^{-3/2} = -3/(8x^{3/2})$$
$$y'''(4) = -3/[8(4^{3/2})] = -3/[8 \times 8] = -3/64$$

Section I, Part B

26. **The correct answer is (D).** In order to integrate tan x, use $\tan x = \dfrac{\sin x}{\cos x}$

$$\int_{\pi}^{5\pi/4} \tan x\, dx + \int_{-\pi}^{\pi} \cos x\, dx$$

Using substitution, $u = \cos x$, $du = -\sin x$, and $-du = \sin x$. So, integrate $-du/u = -\ln |u| = -\ln|\cos x|$.

$-\ln|\cos 5\pi/4| - -\ln|\cos \pi| + \sin(\pi) - \sin(-\pi)$

$= -\ln\left|-\dfrac{\sqrt{2}}{2}\right| + \ln|-1| + 0 - 0 = -\ln\left(\dfrac{\sqrt{2}}{2}\right) + \ln 1 = -\ln\left(\dfrac{\sqrt{2}}{2}\right) + 0$

$= -\ln\left(\dfrac{\sqrt{2}}{2}\right) = -(\ln \sqrt{2} - \ln 2) = -\ln \sqrt{2} + \ln 2 = \ln 2 - \ln \sqrt{2}$

27. **The correct answer is (D).** This is the definition of the derivative. The derivative of sec x is sec x tan x.

$\sec(\pi/3) \tan(\pi/3) = (2)(\sqrt{3}) = 2\sqrt{3}$

28. **The correct answer is (C).**

$V = \pi \int (\text{radius})^2 dx$

$V = \pi \int_{4}^{9} (\sqrt{x})^2 dx = \pi \int_{4}^{9} x\, dx = \dfrac{\pi x^2}{2} \Big|_{4}^{9}$

$= \dfrac{\pi}{2}(9^2 - 4^2) = \dfrac{\pi}{2}(81 - 16) = \dfrac{\pi}{2}(65) = \dfrac{65\pi}{2}$

29. **The correct answer is (B).** Following the directions, the definite integral results in $-\cos(t)$ from 0 to x/2, which is $-\cos(x/2) - -\cos(0)$, which is $-\cos(x/2) + \cos 0$, which is $-\cos(x/2) + 1$. Now, differentiate to get a final answer of $\sin(x/2) \times 1/2 + 0$ or $(1/2)\sin(x/2)$.

30. **The correct answer is (C).**

Perimeter = $2l + 2w = 1200$ Area = $l \times w$
$l + w = 600$
$l = 600 - w$ Area = $(600 - w)w = 600w - w^2$
Area′ = $600 - 2w = 0$
$600 = 2w$
$w = 300$ feet

$l = 600 - w = 600 - 300 = 300$ feet
Maximum area = 300´ × 300´ = 90,000 sq. ft.

31. **The correct answer is (D).**

$$\int \frac{xdx}{\sqrt{3+x}}$$
$$u = \sqrt{3+x}$$
$$u^2 = 3+x$$
$$u^2 - 3 = x$$
$$2udu = dx$$

$$2\int \frac{(u^2-3)udu}{u} = 2\int (u^2-3)du$$
$$= 2(u^3/3 - 3u) + C$$
$$= 2u^3/3 - 6u + C$$
$$= (2/3)(3+x)^{3/2} - 6\sqrt{3+x} + C$$
$$= (2/3)\sqrt{3+x}[(3+x) - 9] + C$$
$$= (2/3)\sqrt{3+x}[3+x-9] + C$$
$$= (2/3)\sqrt{3+x}[x-6] + C$$
$$= [2(x-6)(3+x)^{1/2}]/3 + C$$

32. **The correct answer is (D).**

$$6 = e^{5x}$$
$$\ln 6 = \ln e^{5x}$$
$$\ln 6 = 5x$$
$$x = (\ln 6)/5$$

33. **The correct answer is (E).**

$M(t) = -t^2 + 2t + 4$
$M'(t) = -2t + 2$
$M'(-1) = -2(-1) + 2 = 2 + 2 = 4$

$y = mx + b$
$M = mt + b$
$1 = 4(-1) + b$
$1 = -4 + b$

$$5 = b$$
$$M = 4t + 5$$

34. **The correct answer is (E).** The normal line is perpendicular to the tangent line.

$$y = (x+5)/(x-5)$$
$$y' = \frac{(x-5)-(x+5)}{(x-5)^2} = \frac{x-5-x-5}{(x-5)^2} = \frac{-10}{(x-5)^2}$$

Now, when $y = 11$, $x = 6$. (Just solve the equation, letting $y = 11$.)
$$y'(6) = -10/(6-5)^2 = -10/1^2 = -10/1 = -10$$

The normal line, then, has a slope of 1/10.

$$y = mx + b$$
$$11 = (1/10)(6) + b$$
$$11 = 3/5 + b$$
$$b = 52/5$$

$$y = (1/10)x + 52/5$$
$$10y = x + 104$$
$$x - 10y = -104$$

35. **The correct answer is (B).**

$dd/dt = 5$ cm/sec., $dr/dt = 2.5$ cm/sec. $dA/dt = ?$ when $d = 10$ cm and $r = 5$ cm
$A = \pi r^2$
$dA/dt = 2\pi r \, dr/dt$
$dA/dt = 2\pi(5\text{cm})(2.5 \text{ cm/sec.}) = 25\pi \text{ cm}^2/\text{sec.}$

36. **The correct answer is (D).**

$$y = (x+5)(3x) = 3x^2 + 15x$$
$$y' = 6x + 15$$

37. **The correct answer is (E).**

$$f(x) = 3x^2 - 24x + 36$$
$$f'(x) = 6x - 24 = 0$$
$$6x = 24$$
$$x = 4$$

```
         0                    5
←————————|————————————————————|————————→
   f'(0) = -24          f'(5) = 30 - 24 = 6
   Decreasing                Increasing
```

The graph changes direction at $x = 4$.

38. **The correct answer is (C).** Simply substitute t in for x.

$$y = \sqrt{1-x^2}$$
$$y^2 = 1-x^2$$
$$x^2 + y^2 = 1$$

39. **The correct answer is (E).**

$$\lim_{x \to 0} \frac{2\sin x \cos x}{2x} = \lim_{x \to 0} \frac{\sin 2x}{2x} = 1$$

40. **The correct answer is (E).**

$$\lim_{x \to 2} \frac{4x^2 - 16}{x - 2} = \lim_{x \to 2} \frac{4(x^2 - 4)}{x - 2} = \lim_{x \to 2} \frac{4(x-2)(x+2)}{(x-2)} = \lim_{x \to 2} 4(x+2)$$

Section II

1.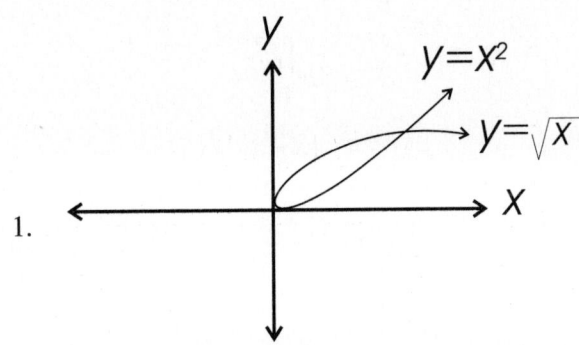

(A) The points of intersection of the two graphs are (0,0) and (1,1).

$$x^2 = \sqrt{x}$$
$$x^4 = x$$
$$x^4 - x = 0$$
$$x(x^3 - 1) = 0$$
$$x = 0 \text{ and } x^3 = 1$$
$$x = 0 \text{ and } x = 1$$

(B) The area is 1/3.

$$\text{Area} = \int \text{top graph} - \text{bottom graph}$$
$$\text{Area} = \int_0^1 \sqrt{x} - x^2 \, dx = [2/3 x^{3/2} - 1/3 x^3] \Big|_0^1$$
$$= [(2/3)(1) - 1/3] - [(2/3)(0) - 0]$$
$$= 2/3 - 1/3 - 0 = 1/3$$

(C) The volume is $3\pi/10$.

Volume $= \pi \int$ (Radius^2 − radius^2)

$$dx = \pi \int_0^1 \left[(\sqrt{x})^2 - (x^2)^2 \right] dx$$

$$= \pi \int_0^1 [x - x^4] dx$$

$$= \pi \left[\frac{x^2}{2} - \frac{x^5}{5} \right]_0^1 = \pi \left[\left(\frac{1}{2} - \frac{1}{5} \right) - (0-0) \right] = \pi \left[\frac{3}{10} \right] = \frac{3\pi}{10}$$

(D) The volume is $3\pi/10$.

Volume $= 2\pi \int$ radius × height $dx =$

$$2\pi \int_0^1 x(\sqrt{x} - x^2) dx = 2\pi \int_0^1 (x^{3/2} - x^3) dx$$

$$= 2\pi[(\frac{2}{5})x^{5/2} - \frac{x^4}{4}]_0^1 = 2\pi[(\frac{2}{5} - \frac{1}{4}) - (0-0)] = 2\pi[\frac{3}{20}] = \frac{3\pi}{10}$$

2.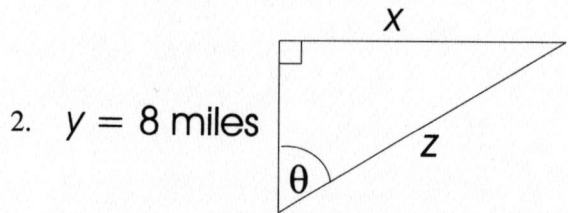
$y = 8$ miles

(A) $d\theta/dt = 12/162 = 2/27$ radians per minute

$dx/dt = -12$ miles per minute $y = 8$ miles
$d\theta/dt = ?$ when $z = 36$ miles

$$\begin{aligned}
\tan \theta &= x/y = x/8 \\
8 \tan \theta &= x \\
8 \sec^2 \theta \, d\theta/dt &= dx/dt \\
8(36/8)^2 \, d\theta/dt &= 12 \\
8(9/2)^2 \, d\theta/dt &= 12 \\
8(81/4) \, d\theta/dt &= 12 \\
162 \, d\theta/dt &= 12 \\
d\theta/dt &= 12/162 = 2/27 \text{ radians per minute}
\end{aligned}$$

PRACTICE TEST 1: ANSWERS AND EXPLANATIONS—SECTION II

(B) $dz/dt = -6\sqrt{3}$ miles per minute

$$dz/dt = ? \text{ when } z = 16 \text{ miles}$$
$$x^2 + y^2 = z^2$$

When $z = 16$, $y = 8$, we have

$$x = \sqrt{z^2 - y^2} = \sqrt{192} = 8\sqrt{3}.$$

$$x^2 + 8^2 = z^2$$
$$2x\, dx/dt + 0 = 2z\, dz/dt$$
$$2(8\sqrt{3})(-12) = 2(16)\, dz/dt$$
$$-96\sqrt{3} = 16\, dz/dt$$
$$dz/dt = -6\sqrt{3} \text{ miles per minute}$$

3. (A) $\quad y = (-11/2)x - 7/2$
$\quad\quad 2y = -11x - 7$
$\quad 11x + 2y = -7$

$$x^3 - 2x^2y + y^2 = -1$$
$$3x^2 - (2x^2 y' + y \times 4x) + 2y\, y' = 0$$
$$3x^2 - 2x^2 y' - 4xy + 2yy' = 0$$
$$(2y - 2x^2)y' = 4xy - 3x^2$$
$$y' = (4xy - 3x^2)/(2y - 2x^2)$$
$$y' \text{ at } (-1, 2) = [4 \times -1 \times 2 - 3(-1)^2] / [2 \times 2 - 2(-1)^2]$$
$$= [-8 - 3]/[4 - 2] = -11/2$$
$$y = mx + b$$
$$2 = (-11/2)(-1) + b$$
$$2 = 11/2 + b$$
$$b = -7/2$$
$$y = (-11/2)x - 7/2$$
$$2y = -11x - 7$$
$$11x + 2y = -7$$

(B) $\quad y = (2/11)x + 24/11$
$\quad\quad 11y = 2x + 24$
$\quad 2x - 11y = -24$

The slope of the normal line at $y = 2$ is $2/11$, (the negative reciprocal of the slope of the tangent line at that point.)

$$2 = (2/11)(-1) + b$$
$$2 = -2/11 + b$$
$$24/11 = b$$

$$y = (2/11)x + 24/11$$
$$11y = 2x + 24$$
$$2x - 11y = -24$$

(C) The tangent line is vertical when $y = x^2$.

The tangent line is vertical when the slope is undefined. The slope is undefined when the denominator is equal to 0.

4. $f(x) = 4x^3 + 6x^2 - 5$

(A) Relative maximum: $(-1, -3)$
Relative minimum: $(0, -5)$

$$f'(x) = 12x^2 + 12x = 0$$
$$12x(x+1) = 0$$
$$x = 0, x = -1$$

```
         -2         -½          1
    <----|----------|----------|---->
              -1          0
     f'(-2) = 24   f'(-½) = -3   f'(1) = 24
     Increasing    Decreasing    Increasing
```

PRACTICE TEST 1: ANSWERS AND EXPLANATIONS—SECTION II

(B) The inflection point is (–1/2, –4).

$$f''(x) = 24x + 12 = 0$$
$$24x = -12$$
$$x = -1/2$$

```
<----|---------|---------|---->
    -1   -½    0
    f"(-1) = -12      f"(0) = 12
    Concave down      Concave up
```

(C) $y = 24x + 35$

$f'(-2) = 24$

$$y = mx + b$$
$$-13 = 24(-2) + b$$
$$-13 = -48 + b$$
$$b = 35$$

(D)
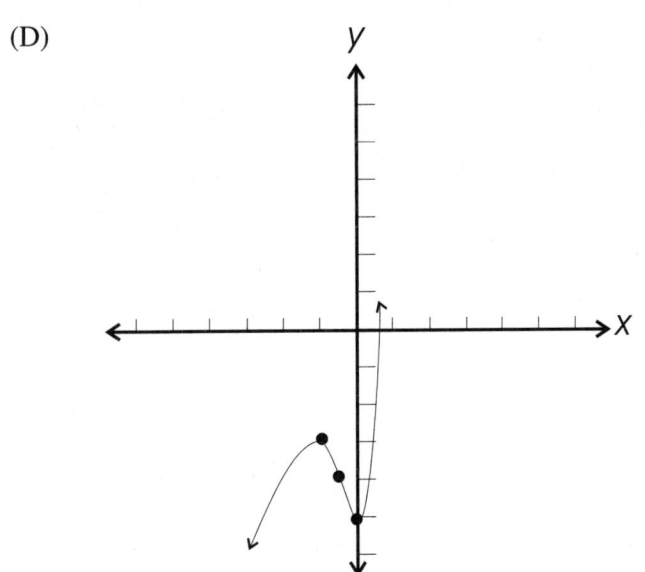

5. $dx/dt = 2$ feet per second

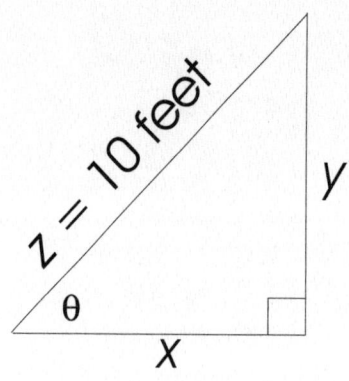

(A) $dy/dt = -20/(10\sqrt{3}) = -2/\sqrt{3} = (-2\sqrt{3})/3$ feet per second

$dy/dt = $? when $x = 5$ feet

$x^2 + y^2 = z^2$
$x^2 + y^2 = 10^2$
when $x = 5$, $y = 5\sqrt{3}$
$2x\, dx/dt + 2y\, dy/dt = 0$
$2(5)(2) + 2(5\sqrt{3})\, dy/dt = 0$
$20 + 10\sqrt{3}\, dy/dt = 0$
$dy/dt = -20/(10\sqrt{3}) = -2/\sqrt{3} = (-2\sqrt{3})/3$ feet per second

(B) $d\theta/dt = -2/(5\sqrt{2}) = -\sqrt{2}/5$ radians per second

$\cos\theta = x/z = x/10$
$-\sin\theta\, d\theta/dt = (1/10)\, dx/dt$
$-\sin(\pi/4)\, d\theta/dt = (1/10)(2) = 1/5$
$-\sqrt{2}/2\, d\theta/dt = 1/5$
$d\theta/dt = -2/(5\sqrt{2}) = -\sqrt{2}/5$ radians per second

PRACTICE TEST 1: ANSWERS AND EXPLANATIONS—SECTION II

6. $A(t) = 100\, e^{kt}$

   (A) $A(5) = 100\, e^{[(\ln 3)/6]5} \approx 249.805$ grams

$$300 = 100\, e^{6k}$$
$$3 = e^{6k}$$
$$\ln 3 = \ln e^{6k}$$
$$\ln 3 = 6k$$
$$k = (\ln 3)/6$$

$$A(t) = 100\, e^{[(\ln 3)/6]t}$$
$$A(5) = 100\, e^{[(\ln 3)/6]5} \approx 249.805 \text{ grams}$$

   (B) $t = 6 \ln 2/\ln 3 \approx 3.786$ hours

$$200 = 100\, e^{[(\ln 3)/6]t}$$
$$2 = e^{[(\ln 3)/6]t}$$
$$\ln 2 = \ln\{e^{[(\ln 3)/6]t}\}$$
$$\ln 2 = (\ln 3)t/6$$
$$6 \ln 2 = (\ln 3)t$$
$$t = 6 \ln 2/\ln 3 \approx 3.786 \text{ hours}$$

   (C) $t = 6 \ln 10/\ln 3 \approx 12.575$ hours

$$1000 = 100\, e^{[(\ln 3)/6]t}$$
$$10 = e^{[(\ln 3)/6]t}$$
$$\ln 10 = \ln\{e^{[(\ln 3)/6]t}\}$$
$$\ln 10 = (\ln 3)t/6$$
$$6 \ln 10 = (\ln 3)t$$
$$t = 6 \ln 10/\ln 3 \approx 12.575 \text{ hours}$$

ADVANCED PLACEMENT CALCULUS AB

PRACTICE TEST 2: ANSWERS AND EXPLANATIONS

Section I, Part A

1. **The correct answer is (C).**

   $f(x) = 6x^2 - 4x + 2$
   Since $f(1) = 6(1)^2 - 4(1) + 2 = 4, f^{-1}(4) = 1$

   Remember that the domain of the original function is the range of its inverse. The range of the original function is the domain of the inverse.

2. **The correct answer is (D).**

   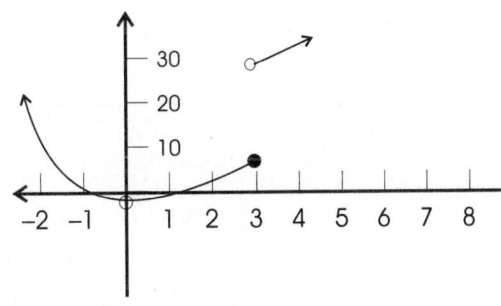

   Look at the graph. See that there is a break only at $x = 3$.

3. **The correct answer is (B).** Setting the denominator $\neq 0$.

   $$x^2 - 25 \neq 0$$
   $$x^2 \neq 25$$
   $$x \neq \pm 5$$

4. **The correct answer is (B).**

   $f(2) = 2^3 + 2(2) = 8 + 4 = 12$

   $g(12) = \sqrt{12-1} = \sqrt{11}$

5. **The correct answer is (D).**

   $C(x) = 12{,}000 + 250x \qquad R(x) = 250x$
   Break-even point: $C(x) = R(x)$

   $12{,}000 + 250x = 350x$
   $12{,}000 = 100x$
   $x = 120$ rugs

6. **The correct answer is (A).**

   $h(x) = 6x^3 - 8x^2 + 2$

   $h'(x) = 18x^2 - 16x$
   $h''(x) = 36x - 16 = 0$
   $36x = 16$
   $x = 16/36 = 4/9$

   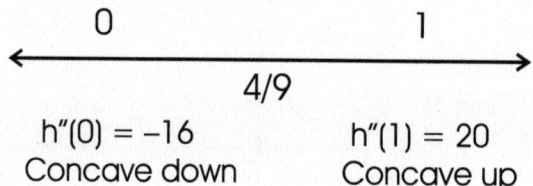

   $h''(0) = -16$ \qquad $h''(1) = 20$
   Concave down \qquad Concave up

   The inflection occurs when $x = 4/9$.

7. **The correct answer is (D).**

   $[1/(4-1)]\int_1^4 e^{2x}\,dx = (1/3)(1/2)e^{2x}\Big|_1^4 = (1/6)[e^{2(4)} - e^{2(1)}]$
   $= (1/6)[e^8 - e^2] = [e^8 - e^2]/6$

8. **The correct answer is (D).**

$$yy' = \frac{1}{x}$$

$$y\, dy/dx = \frac{1}{x}$$
$$y\, dy = dx/x$$
$$y^2/2 = \ln|x| + C_1$$
$$y^2 = 2\ln|x| + C_2$$
$$1^2 = 2\ln(2) + C_2$$
$$1 = \ln 2^2 + C_2$$
$$1 = \ln 4 + C_2$$
$$1 - \ln 4 = C_2$$
$$y^2 = 2\ln|x| + 1 - \ln 4$$

9. **The correct answer is (E).**
$A = P\, e^{rt}$
$A = 1000\, e^{(.0525)6}$
$A = \$1,370.26$

10. **The correct answer is (C).**
$w = (3z^2 - 5z)^5$
$w' = 5(3z^2 - 5z)^4(6z - 5)$

11. **The correct answer is (A).**
$dV/dt = 3$ m³/sec.   radius = 5 m   height = 6 m
$dh/dt = ?$ when radius = 4 m

radius/height = 5/6

$6r = 5h$
$r = (5h)/6$

$$V = (1/3)\pi r^2 h = (1/3)\pi[(5h)/6]$$
$$= (25/108)\pi h^3$$
$$dV/dt = (25/36)\pi h^2 dh/dt$$
$$3 = (25/36)\pi(24/5)^2 dh/dt$$
$$3 = (25/36)\pi(576/25) dh/dt$$
$$3 = 16\pi\, dh/dt$$
$$dh/dt = 3/(16\pi)$$

12. **The correct answer is (C).**

$$y' = (x^3)/(y^2)$$
$$dy/dx = (x^3)/(y^2)$$
$$y^2 dy = x^3 dx$$
$$y^3/3 = x^4/4 + C_1$$
$$y^3 = (3/4)x^4 + C_2$$

13. **The correct answer is (D).**

$4x - 3y = 5 \rightarrow -3y = -4x + 5 \rightarrow y = (4/3)x - (5/3)$, slope = 4/3
slope of perpendicular line = $-3/4$ through point $(-5,4)$

$$y = mx + b$$
$$4 = (-3/4)(-5) + b$$
$$4 = 15/4 + b$$
$$16 = 15 + 4b$$
$$1 = 4b$$
$$b = 1/4$$

14. **The correct answer is (E).**

$$y^2 + 3x - 2y + 8 = 7 - x^2$$
$$x^2 + 3x + y^2 - 2y = 1$$
$$(x + 3/2)^2 + (y - 1)^2 = 1 + 9/4 + 1 = 17/4$$

This is in the form $(x - h)^2 + (y - k)^2 = r^2$, which is a circle.

15. **The correct answer is (C).**
$f(1) = 0$
$g(0) = e^{(3*0)} = e^0 = 1$

16. **The correct answer is (C).**

$$\lim_{x \to 3} \frac{6x^2 - 5}{4x^2 + 1} = \frac{6(3)^2 - 5}{4(3)^2 + 1} = \frac{6(9) - 5}{4(9) + 1} = \frac{54 - 5}{36 + 1} = \frac{49}{37}$$

17. **The correct answer choice is (E).**
$s = -t^3 - 5t$
$s' = -3t^2 - 5 =$ velocity
speed = |velocity|
$s'(7) = -3(7)^2 - 5 = -3(49) - 5 = -147 - 5 = -152 =$ velocity
speed = |–152| = 152

PRACTICE TEST 2: ANSWERS AND EXPLANATIONS—SECTION I, PART A

18. **The correct answer is (D).**

$$\frac{1}{4-1}\int_1^4 (2x+5)^3\,dx$$

$$\frac{1}{3}\int_1^4 (2x+5)^3\,dx$$

$$u = 2x+5$$
$$du = 2dx$$
$$(1/2)du = dx$$

$$\left(\frac{1}{3}\right)\left(\frac{1}{2}\right)\int_7^{13} u^3\,du = \frac{1}{6}u^4/4\Big|_7^{13} = (1/24)\left[13^4 - 7^4\right] = (1/24)[28{,}561 - 2{,}401]$$
$$= (1/24)[26{,}160] = 26{,}160/24 = 1{,}090$$

19. **The correct answer is (B).**

$$f(x) = y = 1/(3x+5)$$
$$x = 1/(3y+5)$$
$$x(3y+5) = 1$$
$$3xy + 5x = 1$$
$$3xy = 1 - 5x$$
$$y = (1-5x)/(3x) = f^{-1}(x)$$

20. **The correct answer is (E).**

$$50y + 25y^2 = 96x - 16x^2 + 231$$
$$16x^2 - 96x + 25y^2 + 50y - 231 = 0$$

Dividing every term by $16 * 25 = 400$

$$x^2/25 - 6x/25 + y^2/16 + y/8 = 231/400$$
$$(x^2 - 6x)/25 + (y^2 + 2y)/16 = 231/400$$

$$\frac{(x-3)^2}{25} + \frac{(y+1)^2}{16} = \frac{231}{400} + \frac{9}{25} + \frac{1}{16} = \frac{231 + 144 + 25}{400} = 1$$

So, this is an ellipse with a center at (3,–1). This graph is not symmetric with any of the choices.

21. **The correct answer is (E).**

$$= \int_2^7 x^3\,dx = \frac{x^4}{4}\Big|_2^7 = \frac{7^4 - 2^4}{4} = \frac{2401 - 16}{4} = \frac{2385}{4}$$

22. **The correct answer is (A).** Take a look at the graph on the calculator!

23. **The correct answer is (A).**

$$y = \frac{x-1}{3+x}$$

$$y' = \frac{(3+x)-(x-1)}{(3+x)^2} = \frac{3+x-x+1}{(3+x)^2} = \frac{4}{(3+x)^2} = 4\cdot(3+x)^{-2}$$

$$y'' = -8(3+x)^{-3}$$

<---------|----------------|--------->
         −4   0
                −3

$y''(-4) = -8(-1)^{-3} = -8/(-1)^3$ $y''(0) = -8(3)^{-3} = -8/(3)^3 = -8/27$
$= (-8)/(-1) = 8$ Concave down
Concave up

24. **The correct answer is (B).**

$$\int_1^2 (5x-3)\,dx \qquad \begin{array}{c} \text{Let } u = 5x-3 \\ du = 5dx \\ (1/5)du = dx \end{array}$$

$$(1/5)\int_2^7 u^2\,du \qquad \text{Remember to change the limits of integration!}$$

$$(1/5)(u^3/3)\,\big|_2^7 = \frac{(7^3-2^3)}{15} = \frac{343-8}{15} = \frac{335}{15} = \frac{67}{3}$$

25. **The correct answer is (D).**

$$\lim_{x\to 5}\frac{x^2-x-20}{x-5} = \lim_{x\to 5}\frac{(x-5)(x+4)}{x-5} = \lim_{x\to 5}(x+4) = 9$$

Section I, Part B

26. **The correct answer is (C).**

$$y = \frac{(x-5)}{(x+3)}$$

$$y' = \frac{(x+3)-(x-5)}{(x+3)^2} = \frac{8}{(x+3)^2}$$

$$y'(-1) = 8/2^2 = 8/4 = 2$$

Since 2 is the slope of the tangent line at the point (–1,–3), –1/2 is the slope of the normal line at the point(–1,–3). (A normal line is perpendicular to a tangent line. Perpendicular lines have negative reciprocal slopes.)

$$y = mx + b$$
$$-3 = (-1/2)(-1) + b$$
$$-3 = 1/2 + b$$
$$b = -7/2$$

$$y = (-1/2)x - 7/2$$
$$2y = -x - 7$$
$$x + 2y = -7$$

27. **The correct answer is (B).**

$$y = uv$$
$$y' = uv' + vu'$$
$$y'' = uv'' + v'u' + vu'' + u'v' = uv'' + 2u'v' + u''v$$
$$uy'' = u^2v'' + 2uu'v' + uu''v$$

28. **The correct answer is (A).**

$h(t) = -16t^2 + 128t$
The ball hits the ground when the height is 0.
$0 = -16t^2 + 128t$
$0 = -16t(t - 8)$
So, the ball is on the ground at times 0 and 8.

$h'(t) = \text{velocity} = -32t + 128$
$h'(8) = -32(8) + 128 = -256 + 128 = -128$

ADVANCED PLACEMENT CALCULUS AB

29. **The correct answer is (C).**

$f = 3x^4 - 5x^3 - 9x + 2$
$f' = 12x^3 - 15x^2 - 9$
$f'' = 36x^2 - 30x = 0$
$6x(6x - 5) = 0$
$x = 0$ and $x = 5/6$

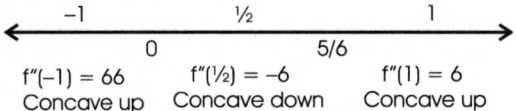

$f''(-1) = 66$ $f''(½) = -6$ $f''(1) = 6$
Concave up   Concave down   Concave up

Concavity changes at inflection points. Hence, there exist two such points.

30. **The correct answer is (E).**

$$\int x^3\left(x - \sqrt{x} + 2\right)dx = \int x^3(x - x^{1/2} + 2)dx = \int x^4 - x^{7/2} + 2x^3 \, dx$$
$$= \frac{x^5}{5} - \frac{2}{9}x^{9/2} + \frac{x^4}{2} + C$$

31. **The correct answer is (E).**

$f(\theta) = 6\cot(2\theta)$
$f'(\theta) = 6 - \csc^2(2\theta) * 2 = -12\csc^2(2\theta)$

32. **The correct answer is (A).**

This is the definition of the derivative.

$$y = \cot(\theta)$$
$$y' = -\csc^2(\theta)$$
$$y'\left(\frac{5\pi}{6}\right) = -\csc^2\left(\frac{5\pi}{6}\right) = -(2)^2 = -4$$

33. **The correct answer is (B).**

$$\int_0^{\pi/4} \sin x \, dx - \int_{-\pi}^{\pi} \cos x \, dx$$

$$-\cos x \Big|_0^{\pi/4} - \sin x \Big|_{-\pi}^{\pi} = \left[-\cos\frac{\pi}{4} + \cos 0\right] - [\sin\pi - \sin(-\pi)]$$
$$= \left[-\sqrt{2}/2 + 1\right] - [0 - 0] = -\sqrt{2}/2 + 1 = \left(-\sqrt{2} + 2\right)/2$$

PRACTICE TEST 2: ANSWERS AND EXPLANATIONS—SECTION I, PART B

34. **The correct answer is (E).**

$$6y = 3e^{2x}$$
$$2y = e^{2x}$$
$$2y' = 2e^{2x}$$
$$y' = e^{2x}$$

35. **The correct answer is (D).**

Arccos[(sin(7π/6)] =
Arccos[−1/2] = (This falls in quadrant II.)
Since π/3 has a cosine of 1/2, then the answer is the angle in quadrant II with a reference angle of π/3. That is 2π/3.

36. **The correct answer is (E).**

$$y = \tan(3x - 5\pi)$$
$$y' = \sec^2(3x - 5\pi) * 3 = 3\sec^2(3x - 5\pi)$$
$$y'(\pi) = 3\sec^2(3*\pi - 5\pi) = 3\sec^2(-2\pi) = 3(1)^2 = 3(1) = 3$$

37. **The correct answer is (B).**

$$dC/dt = 10$$
$$dr/dt = ? \text{ when } d = 5$$

$$C = 2\pi r$$
$$dC/dt = 2\pi \; dr/dt$$
$$10 = 2\pi \; dr/dt$$
$$5/\pi = dr/dt$$

Note that $d = 5$ is irrelevant to this question.

38. **The correct answer is (C).**

A cusp occurs at $x = -3$. The graph is discontinuous at $x = 3$.
A sharp point occurs at $x = 0$.

39. **The correct answer is (B).**

$f(x) = 5x^4 - 2x^3 + 3$
The domain of f is the range of f^{-1}; the range of f is the domain of f^{-1}.

x	$f(x)$	x	$g(x) = f^{-1}(x)$
-1	10	10	-1

$f'(x) = 20x^3 - 6x^2$
$f'(-1) = 20(-1) - 6(1) = -20 - 6 = -26$
$g'(10) = 1/\{f'[g(10)]\} = 1/\{f'[-1]\} = 1/(-26) = -1/26$

40. **The correct answer is (B).**

$y(x) = \ln e^{6 \csc x + 5x}$
$y(x) = 6 \csc x + 5x$
$y'(x) = 6(-\csc x \cot x) + 5$
$y'(x) = -6 \csc x \cot x + 5$

Section II

1. (A) $18,587.48

$$A = Pe^{rt}$$
$$75{,}000 = Pe^{(.0775)(18)}$$
$$75{,}000 = Pe^{1.395}$$
$$P \approx 18{,}587.48$$

(B) 9.300%

$$A = Pe^{rt}$$
$$75{,}000 = 18{,}587.48 e^{r(15)}$$
$$75{,}000/18{,}587.48 = e^{15r}$$
$$\ln(75{,}000/18{,}587.48) = \ln e^{15r} = 15r$$
$$\left[\ln(75{,}000/18{,}587.48)\right]/15 = r$$
$$r \approx 9.300\%$$

(C) 11.409 years

$$A = Pe^{rt}$$
$$45{,}000 = 18{,}587.48 e^{.0775t}$$
$$45{,}000/18{,}587.48 = e^{.0775t}$$
$$\ln(45{,}000/18{,}587.48) = \ln e^{.0775t} = .0775t$$
$$\left[\ln(45{,}000/18{,}587.48)/.0775 = t \approx 11.409 \text{ years}\right]$$

2. (A) $\sqrt{2} - 1$

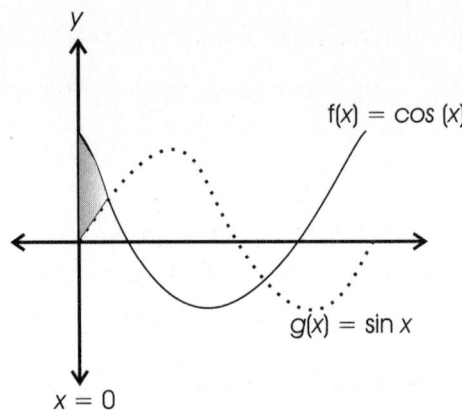

Area = ∫ (top − bottom) dx

$$\int_0^{\pi/4} (\cos x - \sin x)dx = \sin x + \cos x \Big|_0^{\pi/4} = (\sin\frac{\pi}{4} + \cos\frac{\pi}{4}) - (\sin 0 + \cos 0)$$

$$= [\sqrt{2}/2 + \sqrt{2}/2] - [0 + 1]$$

$$= \sqrt{2} - 1$$

(B) $\pi/2$

Volume = $\pi \int [(\text{top radius})^2 - (\text{bottom radius})^2] \, dx =$

$$\pi \int_0^{\pi/4} [(\cos x)^2 - (\sin x)^2] dx$$

$$\pi \int_0^{\pi/4} [\cos 2x] dx = \pi \frac{\sin 2x}{2} \Big|_0^{\pi/4} = \frac{\pi}{2}\{\sin[2\frac{\pi}{4}] - \sin[2(0)]$$
$$= (\pi/2)\{\sin(\pi/2) - \sin(0)\} = (\pi/2)\{1 - 0\} = (\pi/2)\{1\} = \pi/2$$

(C) $\dfrac{\pi^2\sqrt{2}-4\pi}{2}$

Volume $= 2\pi \int \text{radius(height)}\, dx = 2\pi \int_0^{\pi/4} x(\cos x - \sin x)\,dx$

Using integration by parts, this becomes

$= 2\pi\,[x(\sin x + \cos x) - 1(-\cos x + \sin x)]\,\big|_0^{\pi/4}$

$= 2\pi[x\sin x + x\cos x + \cos x - \sin x]\,\big|_0^{\pi/4}$

$= 2\pi\{[(\pi/4)\sin(\pi/4)+(\pi/4)\cos(\pi/4)+\cos(\pi/4)-\sin(\pi/4)] - [0+0+\cos(0)-\sin(0)]\}$

$= 2\pi\{[(\pi/4)(\sqrt{2}/2) + (\pi/4)(\sqrt{2}/2) + \sqrt{2}/2 - \sqrt{2}/2] - [1]\}$

$= 2\pi\{[\pi\sqrt{2}/8 + \pi\sqrt{2}/8] - 1\}$

$= 2\pi\{\pi\sqrt{2}/4 - 1\}$

$= \pi 2\sqrt{2}/2 - 2\pi$

$= (\pi 2\sqrt{2} - 4\pi)/2$

(D) $(\pi^2\sqrt{2} + 4\pi\sqrt{2} - 8\pi)/2$

Volume $= 2\pi \int \text{radius(height)}\, dx = 2\pi \int_0^{\pi/4} (x+1)(\cos x - \sin x)\,dx$

Using integration by parts, this is

$= 2\pi\,[(x+1)(\sin x + \cos x) - 1(-\cos x + \sin x)]\,\big|_0^{\pi/4}$

$= 2\pi[x\sin x + x\cos x + \sin x + \cos x + \cos x - \sin x]\,\big|_0^{\pi/4}$

$= 2\pi[x\sin x + x\cos x + 2\cos x]$

$= 2\pi\{[(\pi/4)\sin(\pi/4)+(\pi/4)\cos(\pi/4)+2\cos(\pi/4)] - [0+0+2\cos(0)]\}$

$= 2\pi\{[(\pi/4)(\sqrt{2}/2) + (\pi/4)(\sqrt{2}/2) + 2(\sqrt{2}/2)] - [2(1)]\} =$

$= 2\pi\{\pi\sqrt{2}/8 + \pi\sqrt{2}/8 + \sqrt{2} - 2\}$

$= 2\pi\{\pi\sqrt{2}/4 + \sqrt{2} - 2\}$

$= \pi 2\sqrt{2}/2 + 2\pi\sqrt{2} - 4\pi$

$= (\pi 2\sqrt{2} + 4\pi\sqrt{2} - 8\pi)/2$

3. (A) 0

$$\lim_{x \to 5^+}(x-5) = 5-5 = 0$$

(B) ∞

The $\lim(\ln(1+x))$, with x approaching infinity is $\ln(\infty)$, which is ∞.

(C) 0

$$\lim_{x \to 0^+} \sin x = s\sin 0 = 0.$$

4. $f(0) = 0$ means the point is $(0,0)$
$f(-1) = 1$ means the point is $(-1,1)$
$f'(-1) = 0$ means that there is a horizontal tangent at $x = -1$
$f''(x) > 0$ on $(-\infty,-1)$ means that the curve is concave up when $x < -1$
$f''(x) < 0$ on $(-1,0)$ and $(0,\infty)$ means that the curve is concave down from -1 to 0 and when $x > 0$.
$f'(x) > 0$ for $x > 0$ means that the curve is increasing when $x > 0$

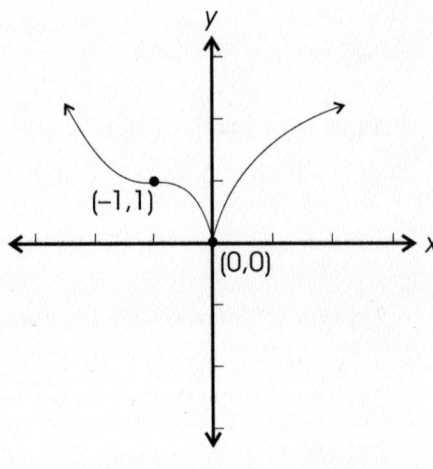

PRACTICE TEST 2: ANSWERS AND EXPLANATIONS—SECTION II

5. $A(2) = 400 \qquad A(6) = 25{,}600$

$$A = Ce^{kt}$$
$$400 = Ce^{k(2)} \qquad 25{,}600 = Ce^{k(6)}$$
$$C = 400/\left(e^{2k}\right) \qquad C = 25{,}600/\left(e^{6k}\right)$$

$$\frac{400}{e^{2k}} = \frac{25{,}600}{e^{6k}}$$
$$400 e^{6k} = 25{,}600 e^{2k}$$
$$e^{6k} = 64 e^{2k}$$
$$\ln e^{6k} = \ln\left(64 e^{2k}\right) = \ln 64 + \ln e^{2k}$$
$$6k = \ln 64 + 2k$$
$$4k = \ln 64$$
$$k = \frac{\ln 64}{4}$$

$$A(2) = C\, e^{[(\ln 64)/4](2)} = 400$$

$$C = \frac{400}{e^{[(\ln 64)/2]}} = 50$$

(A) 50 bacteria

$$A = C\, e^{kt}$$
$$A(0) = 50\, e^{k(0)} = 50\, e^{0} = 50(1) = 50$$

(B) 2/3 hour

$$2C = Ce^{[(\ln 64)/4]t}$$
$$2 = e^{[(\ln 64)/4]t}$$
$$\ln 2 = \ln e^{[(\ln 64)/4]t}$$
$$\ln 2 = [(\ln 64)/4]t$$
$$t = \frac{\ln 2}{\frac{\ln 64}{4}} = 4\frac{\ln 2}{\ln 64} \approx .667$$

(C) 6.644 hours

$$50{,}000 = 50e^{[(\ln 64)/4]t}$$
$$1{,}000 = e^{[(\ln 64)/4]t}$$
$$\ln 1{,}000 = \ln e^{[(\ln 64)/4]t} = [(\ln 64)/4]t$$
$$t = \frac{\ln 1{,}000}{\frac{\ln 64}{4}} = 4\frac{\ln 1{,}000}{\ln 64} \approx 6.644$$

6. $11.50

$$\text{Revenue} = \text{price per ticket} \times \text{number of tickets}$$
$$R = (12 - 1x) \times (11{,}000 + 1{,}000x)$$
$$R = 132{,}000 + 1{,}000x - 1{,}000x^2$$
$$R' = 1{,}000 - 2{,}000x = 0$$
$$2{,}000x = 1{,}000$$
$$x = .50$$
$$\text{Price per ticket} = 12 - 1(.50) = 11.50$$

ADVANCED PLACEMENT CALCULUS BC

PRACTICE TEST 1: ANSWERS AND EXPLANATIONS

Section I, Part A

1. **The correct answer is (A).**

$$\int_1^5 2f(x) - 1\,dx = 2\int_1^5 f(x)\,dx - \int_1^5 1\,dx = 2\times 2 - (5-1) = 0$$

2. **The correct answer is (A).** Making the substitution $u = \sin x$ turns the problem into $\int_0^{\frac{1}{2}} (u^2 + 1)\,du$. The anti-derivative is $\frac{u^3}{3} + u$; evaluating at $\frac{1}{2}$ and at 0 gives the answer.

3. **The correct answer is (B).** The derivative $\frac{dy}{dx}$ is equal to $\frac{dy}{dt}\frac{dt}{dx}$.

    Using the chain rule, $\frac{dy}{dt} = 3e^{3t}$, and $\frac{dt}{dx} = \frac{1}{\left(\frac{dx}{dt}\right)} = \frac{1}{\cos t}$

4. **The correct answer is (D).** Remember that $\lim_{x \to 0} \frac{\sin x}{x} = 1$. Thus

$$\lim_{x \to 0} \frac{\sin^2(3x)}{x^2} = \left(\lim_{x \to 0} \frac{\sin(3x)}{x}\right)^2$$
$$= \left(3\lim_{x \to 0} \frac{\sin(3x)}{3x}\right)^2$$
$$= 9$$

5. **The correct answer is (A).** The function can only have a point of inflection when the second derivative is 0. The derivative is $4x + 2\cos 2x$, so the second derivative is $4 - 4\sin 2x$. This is zero whenever $\sin 2x = 1$, which happens when $x = \dfrac{\pi}{4}$. At this point, the second derivative changes sign, so it is indeed a point of inflection.

6. **The correct answer is (E).** For $f(x)$ to be continuous at $x = 2$, the value of $f(2)$ must be the limit of $f(x)$ as x approaches 2.

$$\lim_{x \to 2} \frac{x^2 - 4}{x - 2} = \lim_{x \to 2} \frac{(x-2)(x+2)}{x-2}$$
$$= \lim_{x \to 2}(x + 2)$$
$$= 4$$

7. **The correct answer is (E).** Multiply through by $\sqrt{x}$ to turn the integral into $\int_1^4 5x^{\frac{3}{2}} + 3x^{\frac{1}{2}} dx = 2x^{\frac{5}{2}} + 2x^{\frac{3}{2}} \Big|_1^4 = 76$.

8. **The correct answer is (C).** Dividing the top and bottom by x^2 turns the limit into $\displaystyle\lim_{x \to \infty} \frac{3\sqrt{1 - \dfrac{3}{x^3}}}{2 + \dfrac{\cos x}{x^2}}$

   Every term involving x goes to zero, so the limit is $\dfrac{3}{2}$.

9. **The correct answer is (B).** Separation of variables gives $\sin y\, dy = x\, dx$. Integrating both sides turns this into $-\cos y = \dfrac{x^2}{2} + c$. Plugging in the given point (2,0) determines that $c = -3$.

PRACTICE TEST 1: ANSWERS AND EXPLANATIONS—SECTION I, PART A

10. **The correct answer is (A).** The substitution $u = 1 - \cos x$ turns the integral in (A) into $\int_0^1 \frac{1}{u} du$, which diverges. The substitution $u = 1 - x$ turns (B) into $\int_0^1 \frac{1}{\sqrt{u}} du$, which converges. The substitution $u = x^2$ turns (C) into $\int_{16}^\infty \frac{1}{2} \cdot e^{-u} du$, which is equal to $\frac{1}{2}\lim_{a \to \infty}(-e^{-a} + e^{-16})$, which in turn is equal to $\frac{e^{-16}}{2}$.

11. **The correct answer is (D).** Use the chain and product rules to determine $f'(x) = \sqrt{2x-1} + \frac{x}{\sqrt{2x-1}}$. Plugging in $x = 5$ gives the answer.

12. **The correct answer is (E).** This is the limit definition of the derivative of $\sin x$ at $x = \frac{\pi}{2}$.

13. **The correct answer is (A).** The area is given by the integral $\int_0^2 xe^{x^2} dx$. Substitute $u = x^2$ to turn this integral into $\int_0^4 \frac{1}{2} e^u du$, which is equal to $\frac{e^4}{2} - \frac{1}{2}$.

14. **The correct answer is (B).** (I) is necessary for the sum to converge, but does not guarantee its convergence. As $\sum_{n=1}^\infty \frac{1}{n^2}$ converges, so does $\sum_{n=0}^\infty f(n)$, as long as $f(x) < \frac{1}{x^2}$, by the comparison test. Knowing that $f(x) > \frac{1}{x^2}$ tells nothing about convergence or divergence of the sum.

15. **The correct answer is (D).** The absolute maximum must occur when $y' = 0$. Using the quotient rule, the derivative is $\dfrac{32-2x^2}{(x^2+16)^2}$, which is equal to 0 when $x = \pm 4$. When $x < -4$ or $x > 4$, y' is negative. When $-4 < x < 4$, y' is positive. Thus the local maximum must occur when $x = 4$ and $y = \dfrac{1}{4}$. As the function has the asymptote $y = 0$, this is the absolute maximum.

16. **The correct answer is (A).** The Intermediate Value Theorem guarantees that $f'(x) = 0$ somewhere between $x = 1$ and $x = 2$, as $f'(x)$ changes sign between those two points. It does not necessarily change sign between $x = 0$ and $x = 1$, or between $x = 2$ and $x = 3$.

17. **The correct answer is (B).** The Mean Value Theorem says that $f'(c) = \dfrac{f(3)-f(-1)}{(3-(-1))}$, for some c between 3 and -1.

18. **The correct answer is (B).** The height of the tree is certainly positive. As the tree is growing, its height is increasing, so the derivative is positive. However, its rate of growth is slowing down, so the second derivative is negative.

19. **The correct answer is (E).** The function is increasing when the derivative $\dfrac{6(9-x^2)}{(x^2+9)^2}$ is positive.

20. **The answer is (B).** As $f(x)$ approaches $\pm\infty$ as x approaches -1, (III) is certainly true. As $f'(x) = \dfrac{1}{(x+1)^2}$, which is never equal to 0, it has no local maxima. The second derivative is $-\dfrac{2}{(x+1)^3}$, which is likewise never 0, so there is no point of inflection.

PRACTICE TEST 1: ANSWERS AND EXPLANATIONS—SECTION I, PART A

21. **The correct answer is (E).** By the ratio test, the series converges when

$$\lim_{n \to \infty} \frac{(x-2)^{n+1}}{(x-2)^n} \frac{n!}{(n+1)!}$$

is less than 1. This limit is equal to

$$\lim_{n \to \infty} \frac{x-2}{n+1}$$

which is equal to 0 for all x. Thus the radius of convergence is infinity.

22. **The correct answer is (C).** The two curves cross when $x = 1$ and when $x = 2$, so those are the limits of integration. The curve $4 - x$ is on top, so the volume is given by $\pi \int_1^2 (4-x)^2 - \left(\frac{6}{x+1}\right)^2 dx$.

23. **The correct answer is (E).** The slope is 0 whenever x or y is equal to 0, so $\frac{dy}{dx}$ cannot be $\frac{x}{y}$ or $\frac{x^2}{y}$. It is non-negative whenever y is positive, whether x is positive or negative, so the answer cannot be xy^2 or xy. This leaves (E) as the only possible answer.

24. **The correct answer is (D).** The value of $g(6)$ is the area under the curve from 0 to 6. This can be calculated using basic geometry, calculating the area of triangles and rectangles.

25. **The correct answer is (B).** When x is between 4 and 5, $f(x)$ is horizontal, so the derivative is 0. When $x = 1$, there is a sharp point, so the derivative is undefined. When $x = 2$, the function is sloping downwards, so the derivative is less than 0.

26. **The correct answer is (A).** Using the method of partial fractions gives

$$\frac{9x+10}{2x^2 - x - 6} = \frac{4}{x-2} + \frac{1}{2x+3}.$$

Integrating this, with the substitutions $u = 2x + 3$ and $v = x - 2$, gives the answer.

27. **The correct answer is (B).** The acceleration is the second derivative of position, which is equal to e^{2-t}. When $t = 2$, this is equal to 1.

28. **The correct answer is (E).** Integrating velocity gives the position $x = \dfrac{t^2}{2} + C$, $y = \dfrac{-\cos(\pi t)}{\pi} + C$. Plug in the initial values to get the specific solution $x = \dfrac{t^2}{2} + 1$, $y = \dfrac{-\cos(\pi t)}{\pi} + 2 + \dfrac{1}{\pi}$. Then plugging in $t = 2$ gives the answer.

Section I, Part B

29. **The correct answer is (C).** Graph the position function, the velocity is zero whenever there is a horizontal tangent line.

30. **The correct answer is (D).** The alternating series test shows that (A) converges. The limit comparison test shows that (B) converges, as $\sum_{n=1}^{\infty} \frac{1}{x^{\frac{3}{2}}}$ does. The ratio test shows that (C) converges.

31. **The correct answer is (E).** Let s be the length of one edge of the cube. Then $\frac{ds}{dt} = 5$, and the surface area, A, is equal to $6s^2$. Thus
$$\frac{dA}{dt} = 12s\frac{ds}{dt} = 600 \text{ when } s = 10.$$

32. **The correct answer is (A).** Use integration by parts, with $u = x + 1$ and $dv = \cos x\, dx$, to turn this integral into $(x+1)\sin x - \int \sin x\, dx$, which is in turn equal to $(x + 1) \sin x + \cos x + C$.

33. **The correct answer is (C).** The curve has radius 0 when $\theta = 0$ and then again when $\theta = \frac{\pi}{3}$, so the area inside one loop is given by
$$\int_0^{\pi/3} \frac{1}{2}(4\sin 3\theta)^2\, d\theta.$$ Evaluate this on your calculator to get the answer.

34. **The correct answer is (B).** A right triangle with one side length x and hypotenuse 7 has the other side length $\sqrt{49-x^2}$. Thus the area is $\frac{x\sqrt{49-x^2}}{2}$. The area is maximal when the derivative of the area is equal to 0. The derivative is $\frac{49-2x^2}{\sqrt{49-x^2}}$ which is equal to zero when $x = 4.95$. The resulting area is 12.25.

35. **The correct answer is (D).** The function has a horizontal tangent line whenever $\frac{dy}{dx} = 0$. Implicit differentiation gives $3x^2 - 3y - 3x\frac{dy}{dx} + 2y\frac{dy}{dx} = 0$. Thus a horizontal tangent line occurs when $\frac{(3x^2 - 3y)}{(3x - 2y)}$, which happens when $x^2 = y$. Plugging x^2 in for y in the original equation turns it into $x^3 - 3x^3 + x^4 = 0$, which factors as $x^3(x - 2) = 0$. Thus the tangent line is horizontal when $x = 0$, $y = 0$ and when $x = 2$, $y = 4$.

36. **The correct answer is (A).** This is a geometric series, it converges whenever $r^2 < 5$.

37. **The correct answer is (A).** As $\frac{dx}{dt} = \cos t$ and $\frac{dy}{dt} = 1 - \sin t$, the formula for arc-length is $\int_0^{\pi/3} \sqrt{\cos^2 t + (1 - \sin t)^2}\, dt$. Evaluate this on your calculator.

38. **The correct answer is (C).** The population at time t is e^{kt} times the initial population, for some value of k. When $t = 3$, $e^{3k} = 2$, so $k = 0.231$. The population has tripled when $e^{0.231t} = 3$; take log of both sides to find the answer.

39. **The correct answer is (A).** Use the product rule and the chain rule.

40. **The correct answer is (A).** The speed is $\sqrt{\left(\frac{dx}{dt}\right)^2 + \left(\frac{dy}{dt}\right)^2}$. Take derivatives, $\frac{dx}{dt} = -\sin 3$, and $\frac{dy}{dt} = \frac{1}{4}$ when $t = 3$.

41. **The correct answer is (B).** The function has a vertical tangent line when the derivative is undefined but tending towards infinity. The derivative is $\frac{2}{3(2x+1)^{\frac{2}{3}}}$ which approaches infinity as x approaches $-\frac{1}{2}$.

PRACTICE TEST 1: ANSWERS AND EXPLANATIONS—SECTION I, PART B

42. **The correct answer is (C).** The value of $f(3.9)$ is approximately $f(4) - 0.1(f'(4))$.

43. **The correct answer is (A).** The Taylor series around 0 of $\frac{1}{1-x}$ is $1 + x + x^2 + x^3 + \ldots$, replace x with $-2x$ to get the Taylor series for $\frac{1}{1+2x}$.

44. **The correct answer is (C).** The position is given by the anti-derivative of the velocity, so the position is $\frac{e^{3t}}{3} - t + C$. Plug in $t = 0$ to determine that $\frac{1}{3} + C = 1$, so $C = \frac{2}{3}$. Thus the position when $t = 2$ is $\frac{e^6}{3} - 2 + \frac{2}{3} = \frac{(e^6 - 4)}{3}$

45. **The answer is (C).** The slope of the tangent line is $y'(x)$, which is equal to $\tan x + x \sec^2 x$. Plugging in $x = \frac{\pi}{4}$ gives the answer.

Section II, Part A

1. (A) The third-degree Taylor polynomial about 0 is $2+3x-\dfrac{x^2}{2}+\dfrac{2x^3}{3}$
Plugging in $x = -0.1$ gives an estimate for $f(-0.1)$ of 1.694.

   (B) To find the Taylor polynomial for $g(x)$, replace x in the polynomial for $f(x)$ with $-x^2$ to get

   $$2-3x^2-\dfrac{x^4}{2}$$

   (C) To find the Taylor polynomial for $h(x)$, integrate the polynomial for $f(t)$.

   $$\int_0^x 2+3t-\dfrac{t^2}{2}\,dt = 2x+\dfrac{3x^2}{2}-\dfrac{x^3}{6}$$

2. (A)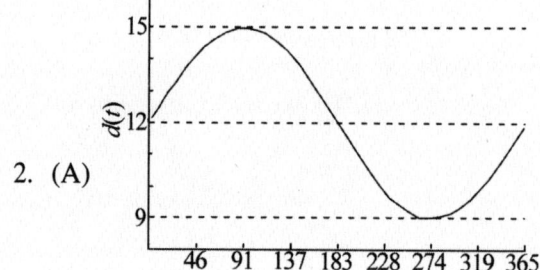

   (B) The average number of hours of sunlight is given by

   $$\int_0^{91} 12+3\sin\left(\dfrac{\pi\,t}{183}\right)dt$$ which is equal to 13.904.

   (C) Excess energy is collected whenever $L(t) > 13$. This occurs when $3\sin\left(\dfrac{\pi\,t}{183}\right) > 1$. As $\sin\left(\dfrac{\pi\,t}{183}\right) > \dfrac{1}{3}$ whenever

   $0.340 < \dfrac{\pi t}{183} < \pi - 0.340$, there is excess energy whenever $19.796 < t < 163.204$.

(D) The amount of money made is

$$0.3\int_{19.796}^{163.204} L(t)-13\, dt$$ which is equal to

$$0.3\int_{19.796}^{163.204} 3\sin\left(\frac{\pi t}{183}\right)-1\, dt$$

This is equal to $55.832.

3. The curve $y=(x+1)^{\frac{1}{2}}$ crosses the line $y = 2$ when $x = 3$.

(A) The area of R is given by $\int_0^3 2-(x+1)^{\frac{1}{2}}\, dx$, which is equal to 4/3.

(B) The volume obtained by rotating R around the x-axis is

$$\pi\int_0^3 2^2-(x+1)\, dx = \pi\left(3x-\frac{x^2}{2}\right)\Big|_0^3 = 14.137$$

(C) We need to find the value of k such that

$$\pi\int_0^k 2^2-(x+1)\, dx = \frac{14.137}{2}$$

Solving the integral turns this into $3k-\frac{k^2}{2}=2.25$. This equation has solutions 0.879 and 5.121. Certainly, the first is the appropriate choice for k. So $k = 0.879$.

4. (A) The position of the particle is equal to $10+\int_0^{20} v(t)\, dt$. Using the left endpoint method to approximate the integral, we get $x(20) \approx 10 + 5(1 + 2 + 7 + 6) = 90$.

(B) The position $x(t)$ increases until t is between 25 and 30. It then starts to decrease, and decreases until $t = 45$, beyond that point it increases again. Thus it is certainly minimal either when $t = 0$ or when $t = 45$. Examining the graph, we see that the area under the velocity graph before it first crosses the x-axis is certainly larger than the area under the graph between its first and second crossing of the x-axis. Thus $x(t)$ does not return to a point smaller than its starting position, so the minimum value of $x(t)$ occurs when $t = 0$.

(C) Acceleration is the derivative of velocity. When $t = 30$, we can approximate this by

$$\frac{v(35) - v(30)}{35 - 30} = -\frac{2}{5}$$

(D) The acceleration is positive whenever the velocity is increasing, this happens when $0 \le t < 15$ and when $35 < t < 50$.

5. (A) To determine the position when $t = 3$, we need to integrate the velocity.

$$x(t) = \int e^{-3t} dt = -\frac{e^{-3t}}{3} + C;$$
plug in $t = 0$, $x(0) = 5/3$ to determine that $C = 2$.

$$y(t) = \int \frac{1}{(t+1)^2} dt = -\frac{1}{t+1} + C;$$
plug in $t = 0$, $y(0) = 2$ to determine that $C = 3$. Thus the position when $t = 3$ is $\left(2 - \frac{e^{-9}}{3}, 3 - \frac{1}{4}\right)$.

(B) The speed is given by $\sqrt{\left(\frac{dx}{dt}\right)^2 + \left(\frac{dy}{dt}\right)^2}$, so the speed when $t = 1$ is

$$\sqrt{e^{-6} + \frac{1}{16}}.$$

(C) As t approaches infinity, $x(t)$ approaches $\lim_{t \to \infty} \left(2 - \frac{e^{-3t}}{3}\right) = 2$ and $y(t)$ approaches $3 - \frac{1}{t+1}$. Thus the particle approaches $(2, 3)$.

6. (A)

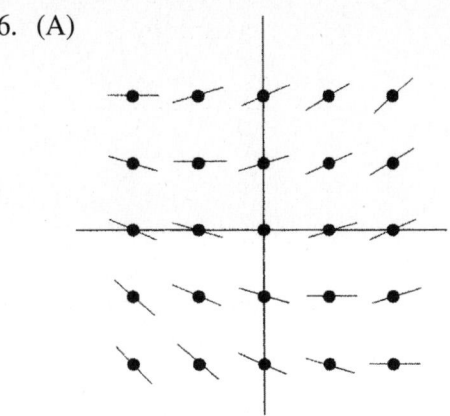

(B) At the point $(0, 4)$, $\frac{dy}{dx} = 1$. Thus $f(0.1)$ is approximately equal to $4 + 0.1 = 4.1$. At the point $(0.1, 4.1)$, $\frac{dy}{dx} = 1.05$, so $f(0.2)$ is approximately equal to $4.1 + (0.1)(1.05) = 4.205$.

(C) Solutions to this differential equation are concave up when the second derivative is positive. The second derivative is

$$\frac{d^2y}{dx^2} = \frac{\left(1 + \frac{dy}{dx}\right)}{4} = \frac{\left(1 + \frac{(x+y)}{4}\right)}{4} = \frac{(4 + x + y)}{16}.$$ This is positive whenever $4 + x + y > 0$, so solutions are concave up at any point where $y > -4 - x$.

ADVANCED PLACEMENT CALCULUS BC

PRACTICE TEST 2: ANSWERS AND EXPLANATIONS

Section I, Part A

1. **The correct answer is (A).** The function is concave down when its second derivative, $e^{1-x^2}(4x^2 - 2)$, is negative. The only one of the five answers that makes this negative is $\frac{1}{2}$.

2. **The correct answer is (D).** As air is being added to the balloon, the radius is certainly positive and increasing. However, as the balloon gets larger, the volume of air needed to increase the radius by one centimeter goes up, so, since air is being added at a constant rate, the rate at which the radius increases goes down. Thus $r''(t)$ is negative.

3. **The correct answer is (E).** The slope is 0 whenever $x = 0$, which means that (D) is not a possibility. It is infinite when $y = 0$, which eliminates (A) and (C). It is negative when x is negative and y is positive, which eliminates (B), this leaves (E) as the only possible answer.

4. **The correct answer is (B).** Use the ratio test:
$$\lim_{n \to \infty} \frac{x^{n+1}}{x^n} \frac{2\ln n}{2\ln(n+1)} = x$$ so the series converges when |x| < 1.

5. **The correct answer is (A).** Separation of variables yields
$\frac{dy}{y} = (3x^2 - 1)dx$. Integrate both sides to get $\ln |y| = x^3 - x + C$. For the solution to pass through the point (2, 1), C must equal −6.

6. **The correct answer is (D).** This is a geometric series, with ratio $\frac{3}{4}$, so its sum is $\frac{1}{1-\frac{3}{4}} = 4$.

7. **The correct answer is (A).** The anti-derivative of acceleration, $(t+c, -\frac{1}{t+1}+k)$, is velocity, when $t = 0$, the velocity is $(0, -1)$, which tells us that $c = 0$ and $k = 0$. The anti-derivative of velocity, $\left(\frac{t^2}{2}+c, -\ln(t+1)+k\right)$, is position. Again, considering the initial value of position tells us that $c = 1$, $k = 3$. Thus, when $t = 2$, the object is at position $(3, 3 - \ln 3)$.

8. **The correct answer is (C).** The function has a local maximum when $f'(x)$ equals zero, and is changing from being positive to being negative.

9. **The correct answer is (C).** $f'(x)$ has positive but decreasing slope on the interval $(0, 1)$, so $f''(x)$ is decreasing there. $f'(x)$ is increasing on $(4, 5)$, so $f''(I)$ is positive there. It is $f(x)$, not $f''(x)$, that has a local minimum when $x = 5$.

10. **The correct answer is (B).** The curves cross when $x = -1, 0, 2$, so the region in question lies between $x = 0$ and $x = 2$. On this interval, the function $y = x + 1$ lies on top of the region. Thus the volume of the region is $\pi \int_0^2 (x+1)^2 - (x^3+1) dx$.

11. **The correct answer is (B).** The function approaches 0 as x approaches infinity, so (C), (D), and (E) are not possibilities. It is never negative, so (A) is eliminated.

12. **The correct answer is (A).** The derivative is $\frac{-4x}{(x^2+4)^2}$, which is 0 when $x = 0$, is positive when $x < 0$ and is negative when $x > 0$, so x is an absolute maximum. The function is defined for all x, and thus has no vertical asymptote, so II is false. The only horizontal asymptote is $y = 0$, so (III) is false.

13. **The correct answer is (D).** By the chain rule $\frac{dy}{dx} = \frac{dy}{dt} \times \frac{dt}{dx}$. $\frac{dy}{dt} = 4\cos 4t$, and $\frac{dx}{dt} = \frac{1}{2\sqrt{t}}$, so $\frac{dt}{dx} = 2\sqrt{t}$.

PRACTICE TEST 2: ANSWERS AND EXPLANATIONS—SECTION I, PART A

14. **The correct answer is (B).** The anti-derivative of velocity is $t^3 - \frac{t^2}{2} + c$, which give the position. As the initial position is 3, $c = 3$. Plugging in $t = 2$ gives the position 9.

15. **The correct answer is (A).** For the function to be continuous at $x = 0$, $f(0)$ must be equal to the limit as x approaches 0 of $f(x)$. Recall that $\lim_{x \to 0} \frac{\sin x}{x} = 1$ so $\lim_{x \to 0} \frac{\sin 2x}{2x} = 1$

    Muliplying both sides by 2 gives the answer.

16. **The correct answer is (C).** Dividing top and bottom by x gives

    $$\lim_{x \to 0} \frac{\sqrt{3 + \frac{2}{x} + \frac{1}{x^2}}}{1 + \frac{1}{x}}$$

    all the terms except the constants approach 0 as x approaches infinity.

17. **The correct answer is (A).** The derivative of $f(x)$ is $e^{x^2-1}(-\sin(x-1) + 2x\cos(x-1))$; plug in $x = 1$ to get the answer.

18. **The correct answer is (A).** The slope of the tangent line is the derivative of the function, which is $-\frac{x}{(x^2 - 5)^{\frac{3}{2}}}$. This is equal to $-\frac{3}{8}$ when $x = 3$.

19. **The correct answer is (E).** The substitution $u = 3x + 1$ turns the integral into $\int_4^7 \frac{1}{3u} du = \frac{(\ln 7 - \ln 4)}{3} = \frac{\ln\left(\frac{7}{4}\right)}{3}$

20. **The correct answer is (C).** The method of partial fractions turns this integral into $\int \frac{2}{x-1} - \frac{1}{3x+1} dx$

21. **The correct answer is (E).** The function is decreasing when the derivative, $\frac{xe^x}{(x+1)^2}$, is negative.

22. **The correct answer is (B).** The minimum value occurs either at an endpoint of the interval or when $f'(x) = 0$. The derivative is $\dfrac{x^2 - 4}{x^2}$, which is 0 when $x = \pm 2$. As $f(1) = 5$, $f(2) = 4$, and $f(3) = \dfrac{13}{3}$, the minimum value is 4.

23. **The correct answer is (C).** Use integration by parts, with $u = x$ and $dv = e^{2x}$, to turn the integral into $\dfrac{xe^{2x}}{2} - \int \dfrac{1}{2}e^{2x}\,dx$, which is equal to $e^{2x}\left(\dfrac{x}{2} - \dfrac{1}{4}\right)$.

24. **The correct answer is (D).** (I) is guaranteed by the Mean Value Theorem, and (III) by the Intermediate Value Theorem. There is no need for (II) to be true.

25. **The correct answer is (A).** The function has a vertical tangent line when the derivative has a vertical asymptote. The derivative
$$f'(x) = \dfrac{e^{\sqrt[3]{x-1}}}{3(x-1)^{\frac{2}{3}}}$$, which has a vertical asymptote when $x = 1$.

26. **The correct answer is (C).** This is the definition of the derivative of the function $f(x) = \sqrt{x}$ when $x = 4$.

27. **The correct answer is (C).** When (I) is the case, the series could either converge or diverge. However, (II) guarantees convergence by the alternating series test, and (III) does so by the comparison test.

28. **The correct answer is (E).** You need more information to determine the integral of the quotient of two functions than just the value of the two integrals.

Section I, Part B

29. **The correct answer is (E).** Adding the first n terms of an alternating series with decreasing terms has an error less than the $n + 1$st term, so you need to add terms until $\frac{1}{2\ln(n+1)} < \frac{1}{10}$. This happens whenever $\ln(n + 1) > 5$, or $n > 147.4$.

30. **The correct answer is (B).** Simplifying shows that $f(x) = \tan x$.

31. **The correct answer is (C).** The speed is the derivative of position, which is $\frac{3\sqrt{t}}{2t^{\frac{3}{2}}+1}$. Plug in $t = 4$ to get the answer.

32. **The correct answer is (C).** An estimate for $f(2.1)$ is given by $f(2) + (0.1)f'(2)$.

33. **The correct answer is (A).** The Taylor series for $\cos x$ is $1 - \frac{x^2}{2} + \frac{x^4}{4!} - \frac{x^6}{6!}.....$; replace x with $\sqrt{x}$ to get the series for $\cos\sqrt{x}$.

34. **The correct answer is (B).** The area under the curve is
$$\int_0^t \sec^2(4x)dx = \frac{1}{4}\tan(4x)\Big|_0^t = \frac{\tan(4t)}{4}.$$

35. **The correct answer is (C).** Divide the top and the bottom by x to get
$$\lim_{x\to\infty} \frac{\frac{2}{x^{\frac{1}{3}}} + \frac{(\sin x)}{x}}{1 - \frac{3}{x}};$$
every term except the constants approaches 0 as x approaches infinity.

36. **The correct answer is (A).** The second derivative of the function is $-\frac{2x^2 + 4x}{(x^2 + 2x + 2)^2}$, which is equal to 0 when $x = 0$ and when $x = -2$

ADVANCED PLACEMENT CALCULUS BC

37. **The correct answer is (D).** When the height of the water is h, the volume of the cone, V, is $\frac{\pi h^3}{12}$. Thus, $\frac{dV}{dt} = \pi \frac{h^2}{4} \frac{dh}{dt}$. When $h = 2$ and $\frac{dV}{dt} = -1$, $\frac{dh}{dt} = \frac{-1}{\pi} = -0.32$

38. **The correct answer is (C).** The series converges when the degree of the denominator is more than 1 greater than the degree of the numerator, by the limit comparison test.

39. **The correct answer is (A).** The area of a rectangle with perimeter 12 is $A(x) = x(6 - x)$. The derivative is $6 - 2x$, which is equal to 0 when $x = 3$. The resulting area is 9.

40. **The correct answer is (C).** The anti-derivative of $\frac{1}{1-x}$ is $-\ln(1 - x)$, which approaches ∞ as x approaches 1, so (A) diverges. The antiderivative of $\sin x$ is $-\cos x$, which has no limit as x approaches ∞, so (B) diverges. The value of (C), on the other hand, is
$$\lim_{x \to \infty}(-2e^{-\sqrt{x}}) + 2e^{-1} = 2e^{-1}.$$

41. **The correct answer is (A).** The two curves cross when $\cos \theta = \frac{1}{2}$, which occurs when $\theta = \frac{\pi}{3}$ and when $\theta = \frac{5\pi}{3}$. The formula for the area of polar curves is $\int_{\pi/3}^{5\pi/3} \left(2\cos^2 \theta - \frac{1}{2}\right) d\theta$, plug this into a direct integral computing program on the calculator to get the answer.

42. **The correct answer is (A).** (A) converges by the alternating series test. (B) diverges by the limit comparison test, comparing it to $\sum_{n=0}^{\infty} \frac{1}{2\sqrt{n}}$, and (C) diverges because the individual terms approach infinity, not 0.

PRACTICE TEST 2: ANSWERS AND EXPLANATIONS—SECTION I, PART B

43. **The correct answer is (B).** As $\frac{dx}{dt} = 2t$ and $\frac{dy}{dt} = \frac{-1}{t^2}$, the formula for arc-length tells us that the arc-length is $\int_1^2 \sqrt{4t^2 + \frac{1}{t^4}}\, dt$

    plug this into a direct integral computing program on the calculator to get the answer.

44. **The correct answer is (D).** The fraction of substance left after t minutes is e^{kt} for some k. When $t = 20$, $e^{20k} = \frac{1}{2}$, so $k = 0.035$. Now plug $t = 11$ in to get the amount left after 11 minutes.

45. **The correct answer is (B).** Speed is $\sqrt{\left(\frac{dx}{dt}\right)^2 + \left(\frac{dy}{dt}\right)^2}$. Calculate $\frac{dx}{dt} = \frac{1}{2\sqrt{t+1}}$, which is equal to $\frac{1}{2\sqrt{3}}$ when $t = 2$, and $\frac{dy}{dt} = -e^{1-t}$, which is equal to $-e^{-1}$ when $t = 2$. Plug these values in to the formula to get the speed.

Section II, Part A

1. (A) The rumor is spreading the most quickly when $r(t)$ has a local maximum. The derivative is $r'(t) = -\dfrac{2000(t-10)}{(1+(t-10)^2)^2}$, which is equal to 0 when $t = 10$. It is positive when $t < 10$ and negative when $t > 10$, so $t = 10$ is a local maximum.

(B) The number of people who know the rumor when $t = 40$ is given by the anti-derivative of $r(t)$. Make the substitution $u = x - 10$ to turn $\int r(t)dt$ into $\int \dfrac{1000}{1+u^2} du = 1000 \tan^{-1}(t-10) + c$: When $t = 0$, $1000 \tan^{-1}(-10) + c = 5$, so $c = 1476.128$. Thus the number of people who know the rumor when $t = 40$ is $1000 \tan^{-1}(30) + 1476.128 = 3013.603$.

(C) The number of people who have heard the rumor at time t is $1000 \tan^{-1}(t-10) + 1476.128$. As t approaches infinity, $\tan^{-1}(t-10)$ approaches $\dfrac{\pi}{2}$, so the number of people who have heard the rumor approaches 3046.924.

2. (A) The area of R is given by $\int_0^1 \dfrac{dx}{x+3} = \ln(x+3) \Big|_0^1$, which is equal to 0.288.

(B) The volume of the solid of revolution is given by

$$\pi \int_0^1 \dfrac{dx}{(x+3)^2} = \pi \left(-\dfrac{1}{x+3}\right) \Big|_0^1 = 0.262$$

(C) The volume of the solid of revolution of the region T is

$$\pi \int_0^1 \dfrac{dx}{(kx+3)^2} = \pi \left(-\dfrac{1}{k(kx+3)}\right) \Big|_0^1 \text{ which is equal to } \pi \left(\dfrac{1}{3(3+k)}\right).$$

We want to know when this is equal to 0.524. Solving for k gives $k = -1.002$.

PRACTICE TEST 2: ANSWERS AND EXPLANATIONS—SECTION II, PART A

3. (A) We need to calculate the arc-length of the curve from $(-2, 7)$ to $(0, 1)$. As $\frac{dy}{dx} = 2x - 1$, the arc-length is $\int_{-2}^{0} \sqrt{1 + (2x-1)^2}\, dx$, which can be calculated on the calculator to be 6.378. Dividing by 3 gives 2.126, the time required to get to the point $(0, 1)$.

(B) Differentiating the equation along which the particle moves with respect to t gives $\frac{dy}{dt} = (2x - 1)\frac{dx}{dt}$, so when $x = 1$, $\frac{dy}{dt} = \frac{dx}{dt}$. As the speed is always equal to 3, $\sqrt{\left(\frac{dx}{dt}\right)^2 + \left(\frac{dy}{dt}\right)^2} = 3$. Thus, when $x = 1$, $\left(\frac{dx}{dt}\right)\sqrt{2} = 3$, so $\frac{dx}{dt} = \frac{dy}{dt} = 2.121$. The velocity is (2.121, 2.121).

(C) As x increases from $x = 1$, the function gets steeper and steeper. Thus, since the particle always travels the same distance (3) in one second, it is speeding up in the y direction, and slowing down in the x direction. Thus $\frac{dy}{dt}$ must be increasing, and $\frac{dx}{dt}$ decreasing.

Section II, Part B

4. (A) The acceleration is the derivative of velocity. When $t = 10$, the acceleration is approximately equal to $\dfrac{(v(12) - v(10))}{2} = -2$.

   (B) The average acceleration is $\dfrac{(v(18) - v(0))}{18} = -\dfrac{1}{18}$.

   (C) As the velocity is always positive, the particle is farthest from its starting point at the end, when $t = 18$.

   (D) The total distance traveled by the particle is the area under the velocity curve. This can be estimated by the left-endpoint rule to be $2(1 + 4 + 8 + 9 + 9 + 7 + 3 + 5 + 4) = 100$.

5. (A)

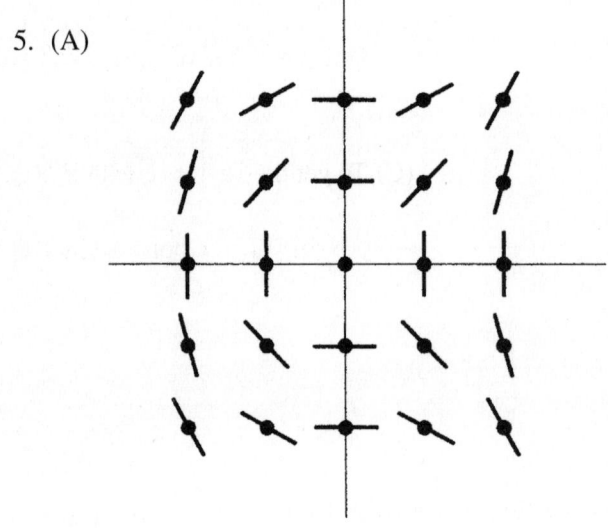

PRACTICE TEST 2: ANSWERS AND EXPLANATIONS—SECTION II, PART B

(B) At the point (0, –1), the slope is 0. Thus, when $x = 0.1$, the curve approximately goes through the point (0.1, –1). The slope at this point is –0.01, so when $x = 0.2$, the curve approximately goes through the point $(0.2, -1 + (0.1)(-0.01)) = (0.2, -1.001)$.

(C) Separating variables gives the equation $ydy = x^2 dx$. Integrate both sides to get $\frac{y^2}{2} = \frac{x^3}{3} + C$. When $x = 0$, $y = -1$, so $C = \frac{1}{2}$. Thus the solution is $\frac{y^2}{2} = \frac{x^3}{3} + \frac{1}{2}$.

6. (A) The Taylor series about 0 for e^x is $1 + x + \frac{x^2}{2} + \frac{x^3}{3!} + \dots$. Replacing x with x^2, we get $1 - x^2 + \frac{x^4}{2}$, the fourth-degree Taylor polynomial for e^{-x^2}.

(B) To get the Taylor series about 0 for $g(x)$, replace x with $2x$ to get 1 $4x^2 + 8x^4$.

(C) To get the Taylor series for $h(x)$, integrate the series for $f(t)$. Thus the fifth degree polynomial is $\int_0^x 1 - t^2 + \frac{t^4}{2} dt = x - \frac{x^3}{3} + \frac{t^5}{10}$

NOTES

NOTES

NOTES

Your everything education destination... the *all-new* Petersons.com

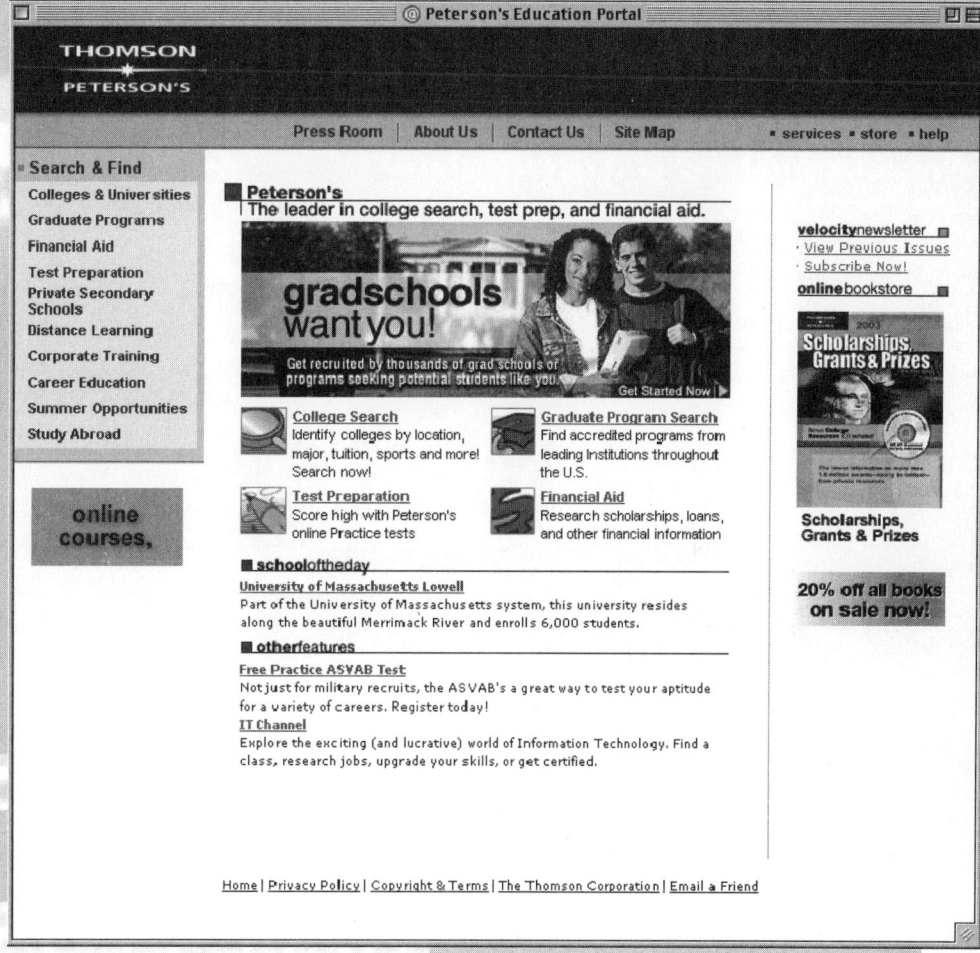

When education is the question, **Petersons.com** is the answer. Log on today and discover what the *all-new* Petersons.com can do for you. Find the ideal college or grad school, take an online practice admission test, or explore financial aid options—all from a name you know and trust, Peterson's.

www.petersons.com

THOMSON
PETERSON'S

Give Your Admissions Essay An Edge At...

EssayEdge.com

Put Harvard-Educated Editors To Work For You

As the largest and most critically acclaimed admissions essay service, EssayEdge.com has assisted countless college, graduate, business, law, and medical program applicants gain acceptance to their first choice schools. With more than 250 Harvard-educated editors on staff, EssayEdge.com provides superior editing and admissions consulting, giving you an edge over hundreds of applicants with comparable academic credentials.

Visit **www.essayedge.com today,** and take your admissions essay to a new level.

"One of the Best Essay Services on the Internet"
—The Washington Post

"The World's Premier Application Essay Editing Service"
—The New York Times Learning Network